Pitman Research Notes in Mathematics Series

Submission of proposals for consideration
Suggestions for publication, in the form of outlines and representative samples, are invited by the Editorial Board for assessment. Intending authors should approach one of the main editors or another member of the Editorial Board, citing the relevant AMS subject classifications. Alternatively, outlines may be sent directly to the publisher's offices. Refereeing is by members of the board and other mathematical authorities in the topic concerned, throughout the world.

Preparation of accepted manuscripts
On acceptance of a proposal, the publisher will supply full instructions for the preparation of manuscripts in a form suitable for direct photo-lithographic reproduction. Specially printed grid sheets are provided and a contribution is offered by the publisher towards the cost of typing. Word processor output, subject to the publisher's approval, is also acceptable.

Illustrations should be prepared by the authors, ready for direct reproduction without further improvement. The use of hand-drawn symbols should be avoided wherever possible, in order to maintain maximum clarity of the text.

The publisher will be pleased to give any guidance necessary during the preparation of a typescript, and will be happy to answer any queries.

Important note
In order to avoid later retyping, intending authors are strongly urged not to begin final preparation of a typescript before receiving the publisher's guidelines and special paper. In this way it is hoped to preserve the uniform appearance of the series.

Longman Scientific & Technical
Longman House
Burnt Mill
Harlow, Essex, UK
(tel (0279) 26721)

Longman Scientific & Technical
Churchill Livingstone Inc.
1560 Broadway
New York, NY 10036, USA
(tel (212) 819-5453)

Titles in this series

1 Improperly posed boundary value problems
A Carasso and A P Stone
2 Lie algebras generated by finite dimensional ideals
I N Stewart
3 Bifurcation problems in nonlinear elasticity
R W Dickey
4 Partial differential equations in the complex domain
D L Colton
5 Quasilinear hyperbolic systems and waves
A Jeffrey
6 Solution of boundary value problems by the method of integral operators
D L Colton
7 Taylor expansions and catastrophes
T Poston and I N Stewart
8 Function theoretic methods in differential equations
R P Gilbert and R J Weinacht
9 Differential topology with a view to applications
D R J Chillingworth
10 Characteristic classes of foliations
H V Pittie
11 Stochastic integration and generalized martingales
A U Kussmaul
12 Zeta-functions: An introduction to algebraic geometry
A D Thomas
13 Explicit *a priori* inequalities with applications to boundary value problems
V G Sigillito
14 Nonlinear diffusion
W E Fitzgibbon III and H F Walker
15 Unsolved problems concerning lattice points
J Hammer
16 Edge-colourings of graphs
S Fiorini and R J Wilson
17 Nonlinear analysis and mechanics: Heriot-Watt Symposium Volume I
R J Knops
18 Actions of fine abelian groups
C Kosniowski
19 Closed graph theorems and webbed spaces
M De Wilde
20 Singular perturbation techniques applied to integro-differential equations
H Grabmüller
21 Retarded functional differential equations: A global point of view
S E A Mohammed
22 Multiparameter spectral theory in Hilbert space
B D Sleeman
24 Mathematical modelling techniques
R Aris
25 Singular points of smooth mappings
C G Gibson
26 Nonlinear evolution equations solvable by the spectral transform
F Calogero
27 Nonlinear analysis and mechanics: Heriot-Watt Symposium Volume II
R J Knops
28 Constructive functional analysis
D S Bridges
29 Elongational flows: Aspects of the behaviour of model elasticoviscous fluids
C J S Petrie
30 Nonlinear analysis and mechanics: Heriot-Watt Symposium Volume III
R J Knops
31 Fractional calculus and integral transforms of generalized functions
A C McBride
32 Complex manifold techniques in theoretical physics
D E Lerner and P D Sommers
33 Hilbert's third problem: scissors congruence
C-H Sah
34 Graph theory and combinatorics
R J Wilson
35 The Tricomi equation with applications to the theory of plane transonic flow
A R Manwell
36 Abstract differential equations
S D Zaidman
37 Advances in twistor theory
L P Hughston and R S Ward
38 Operator theory and functional analysis
I Erdelyi
39 Nonlinear analysis and mechanics: Heriot-Watt Symposium Volume IV
R J Knops
40 Singular systems of differential equations
S L Campbell
41 N-dimensional crystallography
R L E Schwarzenberger
42 Nonlinear partial differential equations in physical problems
D Graffi
43 Shifts and periodicity for right invertible operators
D Przeworska-Rolewicz
44 Rings with chain conditions
A W Chatters and C R Hajarnavis
45 Moduli, deformations and classifications of compact complex manifolds
D Sundararaman
46 Nonlinear problems of analysis in geometry and mechanics
M Atteia, D Bancel and I Gumowski
47 Algorithmic methods in optimal control
W A Gruver and E Sachs
48 Abstract Cauchy problems and functional differential equations
F Kappel and W Schappacher
49 Sequence spaces
W H Ruckle
50 Recent contributions to nonlinear partial differential equations
H Berestycki and H Brezis
51 Subnormal operators
J B Conway

52 Wave propagation in viscoelastic media
F Mainardi

53 Nonlinear partial differential equations and their applications: Collège de France Seminar. Volume I
H Brezis and J L Lions

54 Geometry of Coxeter groups
H Hiller

55 Cusps of Gauss mappings
T Banchoff, T Gaffney and C McCrory

56 An approach to algebraic K-theory
A J Berrick

57 Convex analysis and optimization
J-P Aubin and R B Vintner

58 Convex analysis with applications in the differentiation of convex functions
J R Giles

59 Weak and variational methods for moving boundary problems
C M Elliott and J R Ockendon

60 Nonlinear partial differential equations and their applications: Collège de France Seminar. Volume II
H Brezis and J L Lions

61 Singular systems of differential equations II
S L Campbell

62 Rates of convergence in the central limit theorem
Peter Hall

63 Solution of differential equations by means of one-parameter groups
J M Hill

64 Hankel operators on Hilbert space
S C Power

65 Schrödinger-type operators with continuous spectra
M S P Eastham and H Kalf

66 Recent applications of generalized inverses
S L Campbell

67 Riesz and Fredholm theory in Banach algebra
B A Barnes, G J Murphy, M R F Smyth and T T West

68 Evolution equations and their applications
F Kappel and W Schappacher

69 Generalized solutions of Hamilton-Jacobi equations
P L Lions

70 Nonlinear partial differential equations and their applications: Collège de France Seminar. Volume III
H Brezis and J L Lions

71 Spectral theory and wave operators for the Schrödinger equation
A M Berthier

72 Approximation of Hilbert space operators I
D A Herrero

73 Vector valued Nevanlinna Theory
H J W Ziegler

74 Instability, nonexistence and weighted energy methods in fluid dynamics and related theories
B Straughan

75 Local bifurcation and symmetry
A Vanderbauwhede

76 Clifford analysis
F Brackx, R Delanghe and F Sommen

77 Nonlinear equivalence, reduction of PDEs to ODEs and fast convergent numerical methods
E E Rosinger

78 Free boundary problems, theory and applications. Volume I
A Fasano and M Primicerio

79 Free boundary problems, theory and applications. Volume II
A Fasano and M Primicerio

80 Symplectic geometry
A Crumeyrolle and J Grifone

81 An algorithmic analysis of a communication model with retransmission of flawed messages
D M Lucantoni

82 Geometric games and their applications
W H Ruckle

83 Additive groups of rings
S Feigelstock

84 Nonlinear partial differential equations and their applications: Collège de France Seminar. Volume IV
H Brezis and J L Lions

85 Multiplicative functionals on topological algebras
T Husain

86 Hamilton-Jacobi equations in Hilbert spaces
V Barbu and G Da Prato

87 Harmonic maps with symmetry, harmonic morphisms and deformations of metrics
P Baird

88 Similarity solutions of nonlinear partial differential equations
L Dresner

89 Contributions to nonlinear partial differential equations
C Bardos, A Damlamian, J I Díaz and J Hernández

90 Banach and Hilbert spaces of vector-valued functions
J Burbea and P Masani

91 Control and observation of neutral systems
D Salamon

92 Banach bundles, Banach modules and automorphisms of C*-algebras
M J Dupré and R M Gillette

93 Nonlinear partial differential equations and their applications: Collège de France Seminar. Volume V
H Brezis and J L Lions

94 Computer algebra in applied mathematics: an introduction to MACSYMA
R H Rand

95 Advances in nonlinear waves. Volume I
L Debnath

96 FC-groups
M J Tomkinson

97 Topics in relaxation and ellipsoidal methods
M Akgül

98 Analogue of the group algebra for topological semigroups
H Dzinotyiweyi

99 Stochastic functional differential equations
S E A Mohammed

100 Optimal control of variational inequalities
V Barbu

101 Partial differential equations and dynamical systems
W E Fitzgibbon III

102 Approximation of Hilbert space operators. Volume II
C Apostol, L A Fialkow, D A Herrero and D Voiculescu

103 Nondiscrete induction and iterative processes
V Ptak and F-A Potra

104 Analytic functions – growth aspects
O P Juneja and G P Kapoor

105 Theory of Tikhonov regularization for Fredholm equations of the first kind
C W Groetsch

106 Nonlinear partial differential equations and free boundaries. Volume I
J I Díaz

107 Tight and taut immersions of manifolds
T E Cecil and P J Ryan

108 A layering method for viscous, incompressible L_p flows occupying R^n
A Douglis and E B Fabes

109 Nonlinear partial differential equations and their applications: Collège de France Seminar. Volume VI
H Brezis and J L Lions

110 Finite generalized quadrangles
S E Payne and J A Thas

111 Advances in nonlinear waves. Volume II
L Debnath

112 Topics in several complex variables
E Ramírez de Arellano and D Sundararaman

113 Differential equations, flow invariance and applications
N H Pavel

114 Geometrical combinatorics
F C Holroyd and R J Wilson

115 Generators of strongly continuous semigroups
J A van Casteren

116 Growth of algebras and Gelfand–Kirillov dimension
G R Krause and T H Lenagan

117 Theory of bases and cones
P K Kamthan and M Gupta

118 Linear groups and permutations
A R Camina and E A Whelan

119 General Wiener–Hopf factorization methods
F-O Speck

120 Free boundary problems: applications and theory, Volume III
A Bossavit, A Damlamian and M Fremond

121 Free boundary problems: applications and theory, Volume IV
A Bossavit, A Damlamian and M Fremond

122 Nonlinear partial differential equations and their applications: Collège de France Seminar. Volume VII
H Brezis and J L Lions

123 Geometric methods in operator algebras
H Araki and E G Effros

124 Infinite dimensional analysis–stochastic processes
S Albeverio

125 Ennio de Giorgi Colloquium
P Krée

126 Almost-periodic functions in abstract spaces
S Zaidman

127 Nonlinear variational problems
A Marino, L Modica, S Spagnolo and M Degiovanni

128 Second-order systems of partial differential equations in the plane
L K Hua, W Lin and C-Q Wu

129 Asymptotics of high-order ordinary differential equations
R B Paris and A D Wood

130 Stochastic differential equations
R Wu

131 Differential geometry
L A Cordero

132 Nonlinear differential equations
J K Hale and P Martinez-Amores

133 Approximation theory and applications
S P Singh

134 Near-rings and their links with groups
J D P Meldrum

135 Estimating eigenvalues with *a posteriori/a prior* inequalities
J R Kuttler and V G Sigillito

136 Regular semigroups as extensions
F J Pastijn and M Petrich

137 Representations of rank one Lie groups
D H Collingwood

138 Fractional calculus
G F Roach and A C McBride

139 Hamilton's principle in continuum mechanics
A Bedford

140 Numerical analysis
D F Griffiths and G A Watson

141 Semigroups, theory and applications. Volume
H Brezis, M G Crandall and F Kappel

142 Distribution theorems of L-functions
D Joyner

143 Recent developments in structured continua
D De Kee and P Kaloni

144 Functional analysis and two-point differential operators
J Locker

145 Numerical methods for partial differential equations
S I Hariharan and T H Moulden

146 Completely bounded maps and dilations
V I Paulsen

147 Harmonic analysis on the Heisenberg nilpoten Lie group
W Schempp

148 Contributions to modern calculus of variations
L Cesari

149 Nonlinear parabolic equations: qualitative properties of solutions
L Boccardo and A Tesei

150 From local times to global geometry, control a physics
K D Elworthy

151 A stochastic maximum principle for optimal control of diffusions
U G Haussmann
152 Semigroups, theory and applications. Volume II
H Brezis, M G Crandall and F Kappel
153 A general theory of integration in function spaces
P Muldowney
154 Oakland Conference on partial differential equations and applied mathematics
L R Bragg and J W Dettman
155 Contributions to nonlinear partial differential equations. Volume II
J I Díaz and P L Lions
156 Semigroups of linear operators: an introduction
A C McBride
157 Ordinary and partial differential equations
B D Sleeman and R J Jarvis
158 Hyperbolic equations
F Colombini and M K V Murthy
159 Linear topologies on a ring: an overview
J S Golan
160 Dynamical systems and bifurcation theory
M I Camacho, M J Pacifico and F Takens
161 Branched coverings and algebraic functions
M Namba
162 Perturbation bounds for matrix eigenvalues
R Bhatia
163 Defect minimization in operator equations: theory and applications
R Reemtsen
164 Multidimensional Brownian excursions and potential theory
K Burdzy
165 Viscosity solutions and optimal control
R J Elliott
166 Nonlinear partial differential equations and their applications. Collège de France Seminar. Volume VIII
H Brezis and J L Lions
167 Theory and applications of inverse problems
H Haario
168 Energy stability and convection
G P Galdi and B Straughan
169 Additive groups of rings. Volume II
S Feigelstock

Nonlinear partial differential equations and their applications
Collège de France Seminar
VOLUME VIII

H Brezis & J L Lions (Editors)
Centre National de la Recherche Scientifique / Collège de France
D Cioranescu (Coordinator)

Nonlinear partial differential equations and their applications Collège de France Seminar VOLUME VIII

Copublished in the United States with
John Wiley & Sons, Inc., New York

Longman Scientific & Technical
Longman Group UK Limited
Longman House, Burnt Mill, Harlow
Essex CM20 2JE, England
and Associated Companies throughout the world.

Copublished in the United States with
John Wiley & Sons, Inc., 605 Third Avenue, New York, NY 10158

First published 1988

AMS Subject Classifications: (main) 35, 49, 93
(subsidiary) 65, 73, 76

ISSN 0269-3674

British Library Cataloguing in Publication Data
Nonlinear partial differential equations and
their applications : Collège de France
seminar.—(Pitman research notes in
mathematics series, ISSN 0269–3674; 166).
Vol. 8
1. Differential equations, Partial
2. Differential equations, Nonlinear
I. Brezis, H. II. Lions, J. H. III. Collège
de France
515.3′53 QA377
ISBN 0-582-01605-3

Library of Congress Cataloging-in-Publication Data
(Revised for vol. 8)
Nonlinear partial differential equations and their
applications.
(Research notes in mathematics ;)
Lectures presented at the weekly Seminar on
Applied Mathematics, Collège de France, Paris.
English and French.
Includes bibliographical references.
1. Differential equations, Partial—Congresses.
2. Differential equations, Nonlinear—Congresses.
I. Brezis, H. (Haim) II. Lions, Jacques Louis.
III. Seminar on Applied Mathematics. IV. Series.
QA377.N67 1982 515.3′53 81-4350
ISBN 0-273-08491-7 (v. 1)
ISBN 0-273-08541-7 (v. 2)
ISBN 0-470-20993-3 (v. 8) (USA only)

Printed and bound in Great Britain by
Biddles Ltd, Guildford and King's Lynn

Contents

Preface

The present volume consists of written versions of lectures held during the year 1984-1985 at the weekly Seminar on Applied Mathematics at the Collège de France. They deal mostly with various aspects of the theory of nonlinear partial differential equations.

We thank :

- the speakers who have kindly written up their lectures.
- Mrs Doina Cioranescu who has coordinated the activities of the Seminar and prepared the material for publication; without her patience and determination this volume would never have appeared.

The Seminar is partially supported by a Grant from the C.N.R.S.

Haim BREZIS Jacques Louis LIONS

Préface

Ce volume regroupe les textes des conférences données en 1984-1985 au Séminaire de Mathématiques Appliquées qui se réunit chaque semaine au Collège de France. Elles concernent principalement l'étude d'équations aux dérivées partielles non linéaires sous des éclairages variés.

Nous remercions vivement :

- les conférenciers qui ont bien voulu accepter de rédiger leurs exposés,
- Mme Doina Cioranescu qui s'est chargée de coordonner les activités du Séminaire et de la préparation matérielle de cet ouvrage; sans sa patience et sa persévérance cette publication n'aurait pas vu le jour.

Le séminaire était subventionné en partie par les crédits d'une RCP du C.N.R.S.

Haim BREZIS Jacques Louis LIONS

List of contributors

I.J. Bakelman	Texas A & M University, College Station, Texas 77843, U.S.A. and Institue for Advanced Study, Princeton, NJ 08540, U.S.A.
L. Barthélemy	Equipe de Mathématique de Besançon Université de Franche-Comté, 25030 Besançon Cedex France.
P. Bénilan	Equipe de Mathématique de Besançon Université de Franche-Comté, 25030 Besançon Cedex France.
J.F. Bonnans	INRIA, Domaine de Voluceau, BP 105, Rocquencourt, 78153 Le Chesnay Cedex, France.
E. Casas	Departamento de Ecuaciones Funcionales, Facultad de Ciencias, Universidad de Santander, 39005 Santander, Spain.
F.H. Clarke	Centre de Recherches Mathématiques, Université de Montréal, Case Postale 6128, Succursale A, Montréal, Quebec H3C 3J7, Canada.
A. Haraux	Analyse Numérique, Tour 55-65, 5e étage, Université Pierre et Marie Curie, 4 Place Jussieu, 75230 Paris Cedex 05, France.
D. Jerison	Department of Mathematics, Massachusetts Institute of Technology, Cambridge, MA 02139, U.S.A.
C.E. Kenig	School of Mathematics, University of Minnesota, Minneapolis, MN 55455, U.S.A.
V. Komornik	Eötvös Loránd University, Department of Analysis, H-1088 Budapest, Múzeum Krt. 6-8, Hungary.
O.A. Oleinik	Moscow University, MGU 78, 'K' Kvartal 133, Moscow 117234, Soviet Union.

A.S. Shamaev	Moscow University, MGU 78, 'K' Kvartal 133, Moscow 117234, Soviet Union.
R. Tahraoui	Université de Paris-Sud, Centre Universitaire d'Orsay, Departement de Mathématiques, Bâtiment 425, 91405 Orsay, France.
M.I. Weinstein	Princeton University, Department of Mathematics, Fine Hall, Washington Road, Princeton, NJ08544 U.S.A.
G.A. Yosifian	Moscow University, MGU 78, 'K' Kvartal 133, Moscow 117234, Soviet Union.

I J BAKELMAN

Boundary value problems for n-dimensional Monge-Ampère equations and n-dimensional plasticity equations I

INTRODUCTION

In this paper we are concerned with the boundary value problems for n-dimensional Monge-Ampère and n-dimensional plasticity equations and with related problems of the theory of convex hypersurfaces and functions.

The main topic of Chapter I is the uniform convergence of convex functions in a closed bounded convex n-domain of Euclidean n-dimensional space. We establish necessary and sufficient conditions of such convergence in closed bounded convex domains, whose boundaries are general closed convex hypersurfaces with τ-parabolic support (see definitions in §1).* Here $\tau \geqslant 0$ is a parameter showing the order of parabolic support. If $\tau=0$, then such a convex domain is called strictly convex.** If $\tau>0$, then the class of bounded convex domains with τ-parabolic support is significantly wider than the class of strictly convex domains. For example, it includes convex domains whose boundaries belong to class C^2 and their second fundamental form can vanish on some non-empty set of points. On the other hand, simple examples show that there arise irremovable obstacles for the uniform convergence of convex functions in closed bounded convex domains, whose boundaries contain the straight line segments. Therefore, conditions similar to the condition of τ-parabolic support of closed convex domains are necessary, if we want to study the uniform convergence of convex functions in closed bounded convex domains. The convex domains with τ-parabolic support form the more general class of convex domains, where the necessary and sufficient conditions of such uniform convergence of convex functions were established.

Our recent results (see Chapter I of this paper) include as particular cases our previous results obtained together with Guberman (see [2])*** for

---

* We shall call these domains *convex domains with τ-parabolic support.*

** If in addition the boundary of a strictly convex domain is a C^2-surface, then all its principal normal curvatures are strictly positive everywhere.

*** See also the paper by I. Bakelman, I. Guberman, *The Dirichlet problem for Monge-Ampère equations*, Sibirsk. Mathem. Journal, 4,6,(1963), 1280-1220; (I. Guberman was my graduate student in 1962-1965).

two-dimensional convex domains and also results obtained by myself for strictly convex domains (see [2], [3]). The proof of our recent results for the general case of convex n-domains with τ-parabolic support requires a significantly more difficult technique and more complicated considerations than in our previous papers. Here we present the very detailed proof of main results.

Chapter II is devoted to my recent existence Theorems for the Dirichlet problem

$$\det(u_{ij}) = f(x,u,Du) \qquad (☆)$$

$$u\big|_{\partial G} = h(x) \in C(\partial G) \qquad (☆☆)$$

in a bounded convex domain G with τ-parabolic support. We investigate this Dirichlet problem for convex (elliptic) generalized solutions of the Monge-Ampère equation (☆) by different assumptions with respect to the functions $f(x,u,p)$ in $\bar{G}\times R\times R^n$. First we reduce the Dirichlet problem (☆)-(☆☆) to the operator equations in the space $C(\bar{G})$ by means of geometric techniques of the theory of convex bodies and hypersurfaces and then we investigate them by the different global fixed point theorems in Banach Spaces. Note that the properties of this equation depend on assumptions which we impose on the function $f(x,u,p)$.

The proofs of continuity and compactness for operators participating in these equations are essentially based on the main results of Chapter I of the present paper.

In Chapter III* we are concerned with convex generalized solutions of the Monge-Ampère equation

$$\det(u_{ij}) = f(x,u,Du)$$

defined in the whole space R^n. The graph of every such solution is a complete infinite convex hypersurface in $R^{n+1} = R^n\times R$ and its asymptotic behaviour by $|x|\to\infty$ can be described in the terms of the special convex cone,

* Chapters III and IV will appear in *Nonlinear partial differential equations and their applications*. Collège de France Seminar Volume IX, also to be published in the "Pitman Research Notes in Mathematics" Series.

which is called asymptotic (for the graph of this solution).*

The main problem relating to the Monge-Ampère equation (☆) defined in R^n is as follows : "What kind of sufficiently general Assumptions with respect to the functions f(x,u,p) will provide existence and uniqueness of a convex generalized solution of the equation (☆), whose graph has a prescribed convex non-degenerate asymptotic cone K". This problem was mentioned in the list of the interesting unsolved problems in the Alexandrov-Pogorelov lecture on Diff. Geometry Conference in Leningrad, 1958. This problem was solved by Bakelman in the summer of 1984.

In Chapter III we present the solution of this problem in the class of convex generalized solutions. The proofs of uniqueness and regularity of these solutions for strictly positive and sufficiently smooth functions f(x,u,p) will be given in the forthcoming papers. (Note that assumptions of the positivity and smoothness of the function f(x,u,p) are related only to the proof of regularity of generalized solutions.)

Up to July 1984 only two special cases of equation (☆) had been investigated. The first one is established by Alexandrov [18] and is related to existence and uniqueness of a general complete infinite hypersurface with prescribed area of gaussian mapping (considered as a completely additive set function).

The second Theorem was established by Bakelman [19] in 1956 and is related to generalized solutions of the Monge-Ampère equation

$$\det(u_{ij}) = \frac{g(x)}{R(Du)} \tag{☆☆☆}$$

(the particular case $R(Du) = (1+|Du|^2)^{-\frac{n+1}{2}}$ corresponds to Alexandrov's Theorem). The necessary and sufficient condition of solvability of the main problem considered above for the equation (☆☆☆) is as follows

$$\int_{R^n} g(x)\,dx = \int_{\chi_k(R^n)} R(p)\,dp,$$

where $g(x)\geqslant 0$, $g(x)\in L(R^n)$, $R(p)>0$, $R(p)\in L_{loc}(R^n)$ and $\chi_k(R^n)$ is the tangential image of the convex cone K by means of its supporting hyperplanes. Note that the desired solution is defined to within an additive constant.

---

* See the corresponding definitions in §6.

The non-compactness of domain of equation (☆☆☆) and the complicated implicit necessary condition

$$\text{mes}\ \chi_k(R^n) = \int_{R^n} f(x,u(x),Du(x))\,dx$$

for solutions of the general equation (☆) complicate essentially the proof of the existence Theorem. The proof of the recent Bakelman's existence Theorem for equation (☆) in R^n with prescribed asymptotic convex cone K, presented in Chapter III, is based on the construction of a new Monge-Ampère equation in some special introduced Banach space. This construction is based on the properties of the asymptotic behaviour of the function f(x,u,p) and its first derivatives by $|x|\to\infty$ and $|u|\to\infty$. This recent Theorem includes the existence and uniqueness Theorem for the equation (☆☆☆).

Finally in Chapter IV we prove the existence and uniqueness Theorem for strong solutions of the Dirichlet problem for n-dimensional quasilinear elliptic equations arising from the elastic-plastic torsion problems of hardening rods. These solutions belong to the spaces $C^{m+2,a}(\bar{G})$, where $m\geq 1$, $0<a<1$ and G is a bounded simply connected domain in R^n with $C^{m+2,a}$ boundary ∂G. Equations of such class have the form

$$\sum_{i=1}^{n} \frac{\partial}{\partial x_i}(g(T^2)u_i) = H(x,u) \qquad (P.1)$$

and are called the *n-plasticity equations*. Here T = grad u(x) and the function $g(T^2)$ describes the physical properties of rods. In the real elastic-plastic torsion rods problems we deal with the two-dimensional equation

$$\frac{\partial}{\partial x}(g(T^2)\frac{\partial u}{\partial x}) + \frac{\partial}{\partial y}(g(T^2)\frac{\partial u}{\partial y}) = -2w \qquad (P.2)$$

where u(x,y) is the stress potential and w = const is the torsion per unit length. The experiments with pure shear lead to the dependence between the intensivity of the shear strain tensor Γ and the intensivity of the tangential stress T. We shall consider the simple loading of any isotropic material, then the dependence between Γ and T can be represented by the diagram below and $T^2 = u_x^2+u_y^2$. Initially for $\Gamma<\Gamma_0$ the material follows

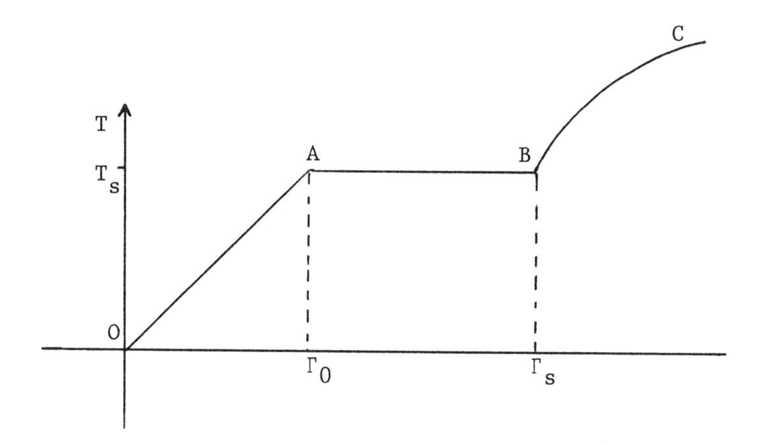

Hooke's law

$$T = G.\Gamma$$

where $G = \text{const}>0$ is called Hooke's constant for the material of rod. This phase from 0 to Γ_0 is called the *linear elastic phase.*

The next phase is pictured by segment AB. This is the yield phase, which is characterized by the growth of shear strain at constant tangential stress T :

$$T = T_s = \text{const}$$

for $\Gamma_0 \leqslant \Gamma \leqslant \Gamma_s$.

The number Γ_s is called the *shear yield limit.* The third part of the diagram corresponding $\Gamma \geqslant \Gamma_s$ is called the *strain-hardening phase*, where the dependence between T and Γ may be written in the form

$$T = f(\Gamma).\Gamma$$

The function f(Γ) is called the modulus of plasticity. Experiments show that

$$0 \leqslant f(\Gamma) \leqslant G$$

for $\Gamma \geqslant \Gamma_s$. In the absence of a yield plateau the strain hardening phase BC joins directly the linear elastic part OA.

Experiments show that $\frac{dT}{d\Gamma}>0$ in the linear elastic and the strain hardening phases. Hence the function $T = f(\Gamma)\Gamma$ is invertible if we consider it as two separate functions correspondingly for $0\leq\Gamma\leq\Gamma_0$ and for $\Gamma\geq\Gamma_s$. Thus we construct two inverses :

$$\Gamma = \frac{1}{G}T \qquad \text{if } 0\leq T\leq T_s$$

and

$$\Gamma = g(T^2)T \quad \text{if } T\geq T_s.$$

The first one corresponds to the linear elastic phase and it converts equation (P.2) to the linear equation

$$\frac{1}{G}\Delta u = -2w$$

and the second one corresponds to the strain-hardening phase with the equation (P.2), which can be non-uniform elliptic.

In this paper we do not consider the applications of different variational methods to the elastic-plastic torsion problems. We note only that the main results obtained by these methods by many authors are presentedin the Friedman monograph [20] (for the torsion problems of linear elastic rods) and the Mikhlin survey [21] (for weak solutions * of the torsion problems of hardening rods).

The main purpose of Chapter IV is the investigation of solvability of the Dirichlet problem in the class of strong solutions for the equation (P.1) in the Hölder spaces $C^{m+1,a}(\bar{G})$, $m\geq1$, $0<a<1$. If we apply our main Theorem 11 (see §8), then we obtain existence and uniqueness Theorems of strong solutions for real torsion problems of hardening rods .

Our main technique is to establish suitable *a priori* estimates for C^1-norms in $\bar{G}$ for smooth solutions of equation (P.1) and then to prove Theorem 11 by the well-known continuity method. In this paper, we develop the new technique based on the one hand on Bakelman's and Serrin's papers [10], [11], [12], [13], [14] devoted to non-uniform elliptic quasilinear

---

* These solutions belong to the Sobolev spaces $W_p^{(1)}(G)$ with $p\geq2$.

equations and hypersurfaces with prescribed mean curvature and on the other hand on the ideas, concepts and facts of the plasticity theory.

We establish the main *a priori* estimates assuming that functions $g(T^2)$, $\Gamma = g(T^2)T$, $\frac{d\Gamma}{dT}$, $\frac{\partial H(x,u)}{\partial u}$ are positive and $g(T^2)$, Γ, $\frac{d\Gamma}{dT}$, $\frac{d^2\Gamma}{dT^2}$ satisfy the suitable general conditions of growth by $T^2 \to +\infty$.

We also assume that some non-trivial integral conditions hold for "coefficients" of equation (P.1) and the boundary of domain G satisfies some local geometric properties (see the corresponding condition of Theorem 11). The last two facts arise from the non-uniform ellipticity of equation (P.1) and the sufficiently arbitrary choice of the function $g(T^2)$.

Clearly the mean curvature equation (see [11], [12], [14]) satisfies Assumptions 12, 13, 14 for the equation (P.1). Therefore the main existence and uniqueness theorems of hypersurfaces with prescribed mean curvature established in [12] follow from Theorem 11 (see Chapter IV of this paper).

Therefore we can consider Theorem 11 as a significant development of existence and uniqueness theorems of smooth hypersurfaces with prescribed mean curvature to the existence and uniqueness theorem for a sufficiently general class of non-uniform elliptic equations, which include interesting non-linear plasticity problems.

The continuation of these investigations will be presented in the forthcoming papers. This paper is divided into two parts; the first one -published in this volume- contains Chapters I and II. The second part -Chapters III and IV- will be published in the next volume of the Seminar.

CHAPTER I. UNIFORM CONVERGENCE OF CONVEX FUNCTIONS IN CLOSED CONVEX DOMAINS

The proofs of the existence theorems of the boundary value problems for elliptic solutions of Monge-Ampère equations are usually based on the uniform convergence of convex functions in a closed bounded convex domain.

If G is a bounded convex domain, then the sufficient conditions of such convergence were obtained in [2], [3] by the exterior sphere assumption regarding ∂G. If ∂G is C^2-hypersurface, then this assumption is equivalent to the inequality

$$\inf_{\partial G} H_i(x) \geq H_0 = \text{const} > 0 \qquad (1)$$

where $H_1(x) \geq H_2(x) \geq \ldots \geq H_{n-1}(x)$ are the principal normal curvatures of ∂G at any point $x \in \partial G$.

We establish that condition (1) can be significantly weakened. We shall prove the theorem of the uniform convergence of convex functions by the assumption, which allows that all principal normal curvatures can vanish at some points of ∂G. We call this assumption the condition of parabolic support of ∂G.

§1. Convex domains with parabolic support

Let G be a convex bounded open domain in R^n and let $\bar{G}$ and ∂G be correspondingly the closure and the boundary of G. If a_o is any point of ∂G, then there exists a supporting (n-1)-plane α of ∂G passing through a_o and an open n-ball $U_\rho(a_o)$ with the center a_o and the radius $\rho > 0$ such that the convex hypersurface

$$\Gamma_\rho(a_o) = \partial G \cap U_\rho(a_o) \tag{1.1}$$

has one-to-one orthogonal projection

$$\pi_\alpha : \Gamma_\rho(a_o) \to \alpha. \tag{1.2}$$

Moreover, the unit normal of α directed to the halfspace of R^n where $\bar{G}$ lies, passes through interior points of G. Clearly all considerations made above also take place for every n-ball $U_{\rho'}(a_o)$ where $0 < \rho' \leq \rho$.

Denote by $\Pi_\rho(a_o)$ the set $\pi_\alpha(\Gamma_\rho(a_o))$ (see figure 1). Let $x_1, x_2, \ldots, x_{n-1}, x_n, z$ be the Cartesian coordinates in R^{n+1} introduced in the following way : a_o is the origin, the axes $x_1, x_2, \ldots, x_{n-1}$ lie in the plane α, the axis x_n is directed along the interior normal of ∂G at the point a_o, and the axis z is orthogonal to the hyperplane R^n (see figure 2). Note that the convex domain G is represented as a three-dimensional in figure 1 and as a two-dimensional in figure 2.

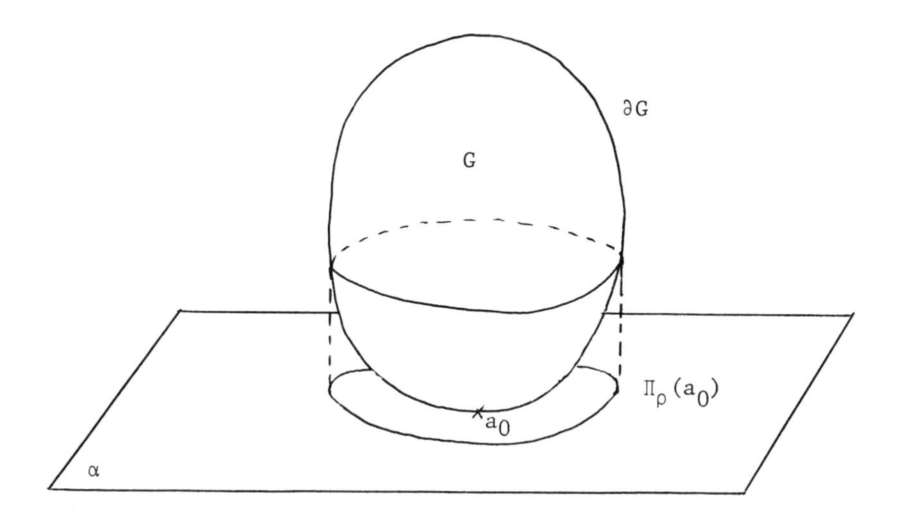

Figure 1

Clearly the convex (n-1)-surface $\Gamma_\rho(a_o)$ is the graph of some convex function

$$\psi(x_1,x_2,\ldots,x_{n-1})\in W^+(\Pi_\rho(a_o)),$$

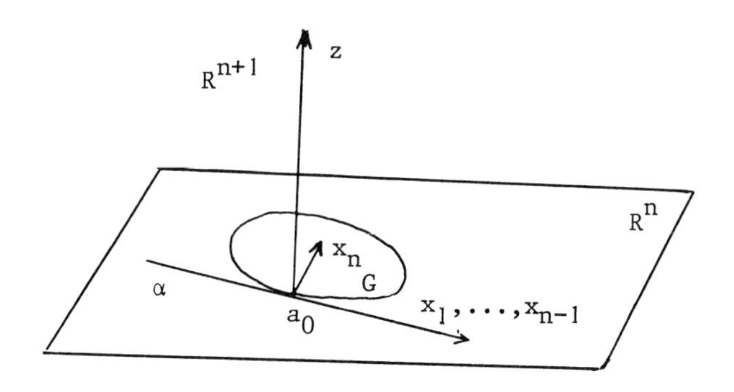

Figure 2

where $W^+(\Pi_\rho(a_o))$ is the set of all convex functions defined in $\Pi_\rho(a_o)$. Obviously,

$$\psi(0,0,\ldots,0) = 0 \tag{1.3}$$

and

$$\psi(x_1,x_2,\ldots,x_{n-1}) \geqslant 0 \tag{1.4}$$

in $\Pi_\rho(a_o)$. The functions $\psi(x_1,x_2,\ldots,x_{n-1})$ *is the local explicit representation of the closed convex surface* ∂G near the marked point a_o. (Note that a_o is an arbitrary point of ∂G).

We shall say that ∂G has *a local parabolic support of order* $\tau \geqslant 0$ at the point a_o if there exist positive numbers ρ_o and $b(a_o)$ such that

$$\psi(x_1,x_2,\ldots,x_{n-1}) \geqslant b(a_o)\left(\sum_{i=1}^{n-1} x_i^2\right)^{2+\tau/2} \tag{1.5}$$

for all $(x_1,x_2,\ldots,x_{n-1}) \in \Pi_{\rho_o}(a_o)$. (Remember that $\rho_o \leqslant \varphi$ and ρ is the radius of the n-ball $U_\rho(a_o)$ which is used in the definition of the convex (n-1)-surface $\Gamma_\rho(a_o)$).

The equivalent formulation of the last concept is as follows : "The convex (n-1)-surface $\Gamma_{\rho_o}(a_o)$ can be touched from the outside by the (n-1)-paraboloid

$$x_n = b(a_o)\left(\sum_{i=1}^{n-1} x_i^2\right)^{2+\tau/2}$$

at the point a_o".

We shall say that ∂G has *a parabolic support of order which is not more than* τ = const$\geqslant 0$, if the local parabolic support of ∂G has an order not more than τ at all points $a_o \in \partial G$.

§2. The statement of the uniform convergence theorem of convex functions in a closed convex domain

2.1 Normal mapping and R-curvature of convex functions. For a detailed exposition see [1], [2, §§16, 17,20], [3].

(A) Normal mapping of convex functions. Let $x_1,x_2,\ldots,x_n,z$ be Cartesian coordinates in R^{n+1} and let R^n be the hyperplane $z = 0$ in R^{n+1}. Let G be an open convex bounded domain in R^n.

We introduce the notations : $x = (x_1, x_2, \ldots, x_n)$ is a point of R^n and $(x,z) = (x_1, x_2, \ldots, x_n; z)$ is a point of R^{n+1}; $z(x)$ is a function $z : G \to R$ with graph S_z; $W^+(G)$ is the set of all convex functions on G; $W^-(G)$ is the set of all concave functions on G. If $z(x) \in W^{\pm}(G)$, then S_z is called a *convex (concave) hypersurface.*

Now pick some arbitrary convex function $z(x) \in W^+(G)$, and let α be a supporting hyperplane of S_z with equation

$$Z - z^\circ = \sum_{i=1}^{n} p_i^\circ (X_i - x_i^\circ)$$

where $(x_1^\circ, \ldots, x_n^\circ, z^\circ) \in S_z \cap \alpha$, and (X,Z) is an arbitrary point of α. We introduce the second copy $\tilde{R}^n$ of the Euclidean space R^n with Cartesian coordinates $p_1, p_2, \ldots, p_n$.

The point $\chi_z(\alpha) = p_o = (p_1^\circ, p_2^\circ, \ldots, p_n^\circ) \in R^n$ is called the normal image of the supporting hyperplane α.

We construct the set

$$\chi_z(x_o) = \bigcup_\alpha \chi_z(\alpha),$$

where α runs through all supporting hyperplanes to S_z at the point $(x_o, z(x_o)) \in S_z$. The set $\chi_z(x_o)$ is called *the normal image of the point* x_o (relative to the function $z(x)$). Clearly $\chi_z(x_o)$ is a closed subset of R^n. Finally, we set for any subset $e \subset G$,

$$\chi_z(e) = \bigcup_{x_o \in e} \chi_z(x_o).$$

<u>The main properties of the normal mapping</u>. These are as follows :

(a) The set $\chi_z(e)$ is a closed subset of $\tilde{R}^n$ for each closed subset e of the domain G; the set $\chi_z(e)$ is a Lebesgue measurable set of R^n for each Borel subset $e \subset G$.

(b) Let $z_1(x)$ and $z_2(x)$ be convex functions, coinciding on G, and $z_1(x) \leq z_2(x)$ for all $x \in G$. Then

$$\chi_{z_2}(G) \subset \chi_{z_1}(G) \tag{2.1}$$

(c) If $z(x)\in W^+(G)\cap C^2(G)$, then the normal mapping can be considered as a mapping of points, namely, the tangential mapping $\chi_z(x) = \text{grad } z(x)$.

(B) R-curvature. Let $R(p)>0$ be a locally summable function on $\tilde{R}^n$. The function of sets

$$\omega(R,z,e) = \int_{\chi_z(e)} R(p)dp, \quad e\subset G \tag{2.2}$$

is non-negative and completely additive on the ring of Borel subsets of the convex domain G for all convex function $z(x)\in W^+(G)$.

This function is called the *R-curvature of the convex function* $z(x)$.

If $R(p) = 1$, then the 1-curvature of $z(x)\in W^+(G)$ is called the area (measure) of the normal mapping and simply denoted by $\omega(z,e)$. If $R(p) = (1+|p|^2)^{-n+1/2}$ then the corresponding R-curvature coincides with the area of the Gauss mapping of the hypersurface S_z. We set

$$A(R) = \int_{R^n} R(p)dp. \tag{2.3}$$

The properties of R-curvature.

(a) Clearly $A(R)>0$; note that the case $A(R) = +\infty$ is not excluded.

(b) The inequality

$$\omega(R,z,G)\leqslant A(R) \tag{2.4}$$

holds for all convex functions $z(x)\in W^+(G)$.

(c) If $z(x)\in W^+(G)\cap C^2(G)$, then

$$\omega(R,z;e) = \int_e \det||z_{ij}||R(\text{grad } z)dx . \tag{2.5}$$

(d) Weak convergence of R-curvature. If the sequence of convex functions $z_n(x)\in W^+(G)$ converges to the convex function $z(x)\in W^+(G)$ in all points $x\in G$, then

$$\lim_{n\to\infty} \int_G \varphi(x)\omega(R,z_n,de) = \int_G \varphi(x)\omega(R,z,de), \tag{2.6}$$

where $\varphi(x)$ is any continuous function in G vanishing outside of some compact subset M of G distant from ∂G on a positive number.

2.2 The border of a convex function. In this subsection we are concerned with the properties of limiting values of convex functions $u(x)\in W^+(G)$, where G is an open bounded convex domain. Let $u(x)\in W^+(G)$. Then u(x) is continuous in G and takes only finite values in G.

Let H_u be the union of points (x,z) where $x\in G$ and $z\geq u(x)$. Then H_u is a convex set and $\bar{H}_u = H_u \cup \partial H_u$ is a convex body in R^{n+1}. Denote by $Z = \partial G\times R$ the cylinder $\{(x,z);x\in\partial G,\ z\in(-\infty,+\infty)\}$. Clearly

$$\partial\bar{H}_u = S_u \cup M_u$$

where S_u is the graph of u(x) in G and M_u is a closed subset of Z such that $M_u \cap S_u = \emptyset$ and the projection of M_u on E^n coincides with ∂G.

We denote by $L_u(x_o)$ the set of all limit points of the convex hypersurface S_u lying on the straight line lx_0. Since S_u is a convex hypersurface and

$$U_o = \inf_G u(x) > -\infty$$

for any function $u(x)\in W^+(G)$, then $L_u(x_o)$ is either some point (x_o,z_o), or some closed segment consisting of points (x_o,z); $z_o\leq z\leq z_1$, or some closed ray consisting of points $\{(x_o,z);z_o\leq z<+\infty\}$ or the empty set.

Now we introduce the function

$$h_u : \partial G\to R$$

by means of the fomula

$$h_u(x_o) = \begin{cases} z_o & \text{if } L_u(x_o) \neq \emptyset; \\ +\infty & \text{if } L_u(x_o) = \emptyset. \end{cases}$$

for all $x_o\in G$. We call the function h_u the border of the function $u(x)\in W^+(G)$.

<u>Remarks</u>. (i) The case $h_u(x_o) = +\infty$ is excluded for any $x_o\in\partial G$, if

$$u(x)\leqslant u_1 = \text{const.} < +\infty$$

for all $x\in G$.

(ii) The function h_u can be discontinuous for some functions $u(x)\in W^+(G)$. Here is the simplest example : consider the convex cone with the base ∂G and the vertex (x_o, z_o), where $x_o\in\partial G$ and $z_o<0$. Then the border of $u(x)$ is the discontinuous function

$$h_u(x) = \begin{cases} 0 & \text{if } x \neq x_o; \\ z_o & \text{if } x = x_o. \end{cases}$$

2.3 The statement of the uniform convergence theorem of convex functions in a closed bounded convex domain. First we present the main assumptions used in the proof of the uniform convergence theorem. These assumptions are very general and natural.

Assumption 1. The boundary ∂G of an open bounded convex domain G has a parabolic support of order not more than $\tau = \text{const}\geqslant 0$.

Assumption 2. The function $R(p)>0$ is locally summable in $\tilde{R}^n$ and the inequality

$$R(p)\geqslant C_o|p|^{-2k}, \tag{2.7}$$

holds for all $p\in\tilde{R}^n$ and $|p|\geqslant r_o$, where $k\geqslant 0$, $r_o>0$ and $C_o>0$ are some constants.

Assumption 3. Let $u_m(x)\in W^+(G)$, $m = 1,2,3,\ldots$ and let $u_m(x)$ converge pointwise to a convex function $u_o(x)\in W^+(G)$.

We assume also that : (i) the borders of $u_m(x)$ are continuous functions $h_{u_m} : \partial G\to R$ and $h_{u_m}(x)$ converge uniformly to some continuous function $h_o : \partial G\to R$ for all $x\in\partial G$; (ii) let x_o be an arbitrary point of ∂G, then there exists some n-ball $U_\rho(x_o)$ such that the inequality

$$\varlimsup_{m\to\infty} \omega(R,u_m,e)\leqslant a[\sup (\text{dist}(x,G))]^{\lambda} \text{ mes } e \tag{2.8}$$

holds for every Borel set $e\subset U_\rho(x_o)\cap G$, where $\lambda = \text{const}\geqslant 0$ and $a = \text{const}>0$.

THEOREM 1. (The uniform convergence theorem of convex functions). Let $u_m(x) \in W^+(G)$, $m=1,2,3,\ldots$ be the sequence of convex functions satisfying the assumption 3. Let the assumption 1 and 2 be also fulfilled. Then

(i) The function $h_o(x)$ is the border of the function $u_o(x) \in W^+(G)$ if either

$$k \leqslant \frac{n+\tau+1}{\tau+2} + \frac{\lambda}{2} \quad \text{for } k<1 \text{ and } k \geqslant \frac{n}{2} \tag{2.9}$$

or

$$k < \frac{n+\tau+1}{\tau+2} + \frac{\lambda}{2} \quad \text{for } 1 \leqslant k < \frac{n}{2} . \tag{2.10}$$

(ii) If we extend the functions $u_m(x)$ and $u_o(x)$ to $\bar{G}$ by means of their borders $h_{u_m}(x)$ and $h_o(x)$, then the extended functions $u_m(x)$ uniformly converge to the extended function $u_o(x)$. □

The proof of this theorem consists of four separate parts which will be developed in the next section.

§3. The proof of Theorem 1

3.1 (Part 1). Suppose that the border $h_{u_o}(x)$ of the convex function $u_o(x)$ does not coincide with the function $h_o(x)$. Since the inequality

$$h_{u_o}(x) \leqslant h_o(x)$$

holds for all $x \in \partial G$ then there exists at least one point $x_o \in \partial G$ such that

$$h_{u_o}(x) < h_o(x_o). \tag{3.1}$$

Now introduce the special Cartesian coordinates in R^n and R^{n+1} with the origin x_o considered above (see figure 2 and corresponding text on page 8, §1). Thus the point $x_o \in R^n$ can be denoted by $0(0,0,\ldots,0)$. Let $Q = (0,h_o(0))$ and $\bar{Q} = (0,h_{u_o}(0))$ be points of the z-axis and let $0<\delta<1$ be an arbitrary number. We introduce two new points $Q'(0,h_{u_o}(0)+\delta\Delta h)$ and $Q''(0,h_{u_o}(0)-\delta\Delta h)$ lying on the z-axis, where

$$\Delta h = h_o(0)-h_{u_o}(0)>0. \tag{3.2}$$

Thus the point Q' lies inside the segment $\overline{QQ}$ and Q" lies under the point $\bar{Q}$. Clearly Δh is the length of the segment Q'Q".

Now consider the hyperplanes β' and β'' in E^{n+1} given by equations :

$$\beta' : z = h_o(0) - \delta\Delta h - \frac{1}{\gamma} x_n; \tag{3.3}$$

$$\beta'' : z = h_{u_o}(0) - \delta\Delta h \tag{3.4}$$

where γ is a sufficient small positive number.

Let $Z = G\times R$ be a cylinder in R^{n+1} with the base ∂G, generators parallel to the z-axis. Then Z bounds some convex body K (figure 3) together with the hyperplane β' and β''.

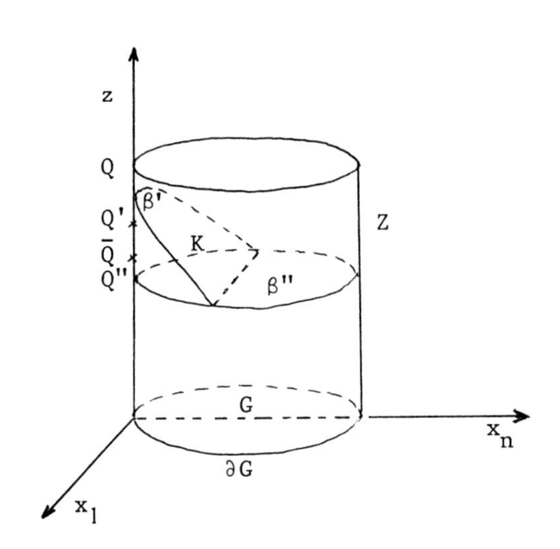

Figure 3

Clearly the convex hypersurfaces S_{u_m} have non-empty intersections with the convex body K. Remember that S_{u_m} is the graph of the convex function $u_m(x)$ (see section 2.1).

We denote by Q_m the nearest point of S_{u_m} to the point $\bar{Q}$ and introduce the sets

$$S_m(K) = S_{u_m} \cap K \tag{3.5}$$

and

$$\beta'(K) = \beta' \cap K. \tag{3.6}$$

If $\gamma > 0$ is sufficiently small and m is sufficiently large then

$$S_m(K) \cap \beta'' \neq \emptyset$$

and

$$S_m(K) \cap Z = \emptyset.$$

Therefore $Q_m \in S_m(K)$. Let $H_m(K)$ be the projection of $S_m(K)$ on R^n and let V_m be the convex cone with the vertex Q_m and the base $\beta'(K)$. Then the normal image of the set $S_m(K)$ covers the normal image of the cone V_m. Therefore

$$\omega(R, V_m) \leqslant \omega(R, u_m, S_m(K)). \tag{3.7}$$

Let $H(K)$ be the projection of the convex body K on R^n. Since

$$H_m(K) \subset H(K),$$

then

$$\omega(R, u_m, H_m(K)) \leqslant \omega(R, u_m, H(K)).$$

Thus

$$\omega(R, V_m) \leqslant \omega(R, u_m, H(K)). \tag{3.8}$$

If $m \to +\infty$, then the points Q_m converge to the point $\bar{Q}$. Hence

$$\lim_{m \to \infty} \omega(R, V_m) = \omega(R, V) \tag{3.9}$$

where V is the convex cone with the vertex $\bar{Q}$ and the base $\beta'(K)$.

Now we use the assumption 3 (see section 2.3), i.e. there exists some n-ball $U_\rho(0)$ for which the inequality (2.8) holds. If $\gamma>0$ is sufficiently small then the Borel set

$$H_m(K)\subset U_\rho(0)\cap G.$$

Hence

$$\underline{\lim}_{m\to\infty}\, \omega(R,u_m,H(K)) \leqslant a[\sup_{H(K)} [(\text{dist}(x,\partial G))]^{\lambda}] \text{ mes } H(K).$$

The last inequality, together with (3.8) implies the inequality

$$\omega(R,V)\leqslant a[\sup_{H(K)}[\text{ dist}(x,\partial G)]^{\lambda}] \text{ mes } H(K). \tag{3.10}$$

Clearly

$$\sup_{H(K)} \text{dist}(x,\partial G) = \gamma.\Delta h$$

Denote by T the set of points $(x_1,x_2,\ldots,x_n)\in R^n$ satisfying the conditions

$$b(0)(\sum_{i=1}^{n-1} x_i^2)^{\frac{2+\lambda}{2}} \leqslant x_n\leqslant\gamma\Delta h$$

where $b(0) = \text{const}>0$ (see the definition of the order of the parabolic support for ∂G; §1). Then $H(K)\subset T$. Now

$$\text{mes } T = \int_T dx = \mu_{n-1} \int_0^{\gamma\Delta h} (\frac{h}{b(0)})^{\frac{n-1}{2+\tau}} dh = d_1\gamma^{\frac{n+\tau+1}{\tau+2}}, \tag{3.11}$$

where $d_1 = \text{const}>0$, depends only on given constants $\tau>0$, $b(0)>0$, $\Delta h>0$ and μ_{n-1}, where Δh is the length of the segment $Q\bar{Q}$ and μ_{n-1} is the volume of the unit (n-1)-ball.

Now from (3.11) and (3.10) it follows that

$$\omega(R,V)\leqslant d_2\gamma^{\lambda+\frac{n+\tau+1}{\tau+2}} \tag{3.12}$$

where $d_2=d_1\, a(\Delta h)^{\lambda} = \text{const}>0$.

The estimate (3.12) completes the first part of the proof of Theorem 1.

3.2 (Part 2). The second part is devoted to obtaining the estimate for $\omega(R,V)$ from below. Let $\tilde{Z} = S_o \times R$ be the cylinder with the base S_o and generators parallel to the z-axis, where S_o is the graph of the equation

$$x_n = b(x_o)\left(\sum_{i=1}^{n-1} x_i^2\right)^{\frac{\tau+2}{2}}$$

in the hyperplane R^n.

Now replace the cylinder Z by the cylinder $\tilde{Z}$ in the constructions mentioned above. Then we obtain the convex one $\tilde{V}$ instead of the convex cone V.

Clearly

$$\omega(R,V) \geqslant \omega(R,\tilde{V}). \qquad (3.13)$$

The estimate $\omega(R,\tilde{V})$ from below is based on the following lemma.

LEMMA 1. Let H^n be the cone of revolution with the axis p_n, the vertex $(0,0,\ldots,0,-\frac{\delta}{\gamma})$ and the base U^{n-1} in the space R^n, where U^{n-1} is the (n-1)-dimensional ball

$$p_1^2+p_2^2+\ldots+p_{n-1}^2 \leqslant (C')^{2-\frac{2}{\tau+2}}$$

$$p_n = -C''\gamma^{-1}.$$

If

$$C' = \frac{\tau+2}{\tau+1}(1-\delta)(\Delta h)^{\frac{\tau+1}{\tau+2}}\, b(0)^{\frac{1}{\tau+2}} \qquad (3.14)$$

$$C'' = \frac{\tau+2-\delta}{\tau+1}, \qquad (3.15)$$

then the normal image of V contains the cone H^n.

PROOF. Let $p_1,p_2,\ldots,p_n$ be the Cartesian coordinates in the space $\tilde{R}^n$ associated with the Cartesian coordinates $x_1,x_2,\ldots,x_n$ with the origin x_o and the axes $x_1,x_2,\ldots,x_{n-1}$ in the (n-1)-supporting plane α of ∂G (see the beginning of section 3.1). Note that the normal image of the convex cone $\tilde{V}$ is some convex closed set in the space $\tilde{R}^n$. Therefore it suffices to prove two following facts :

a) the point $(0,0,0,\ldots,0,-\frac{\delta}{\gamma})$ belongs to the normal image of $\tilde{V}$;

b) the normal image of $\tilde{V}$ contains the (n-1)-ball U^{n-1} mentioned in the statement of the present lemma.

Let β''' be a hyperplane passing through the point $\bar{Q}$ and the (n-1)-plane $\beta'\cap\beta''$. Clearly $\chi(\beta''')\in\chi(\tilde{V})$, where $\chi(\tilde{V})$ is the normal image of the cone $\tilde{V}$. The equation of β''' is

$$z = h_{u_o}(0)-\frac{\delta}{\gamma}x_n.$$

Therefore the point $\gamma(\beta''') = (0,0,\ldots,0,-\frac{\delta}{\gamma})$ belongs to the normal image $\chi(\tilde{V})$ of the convex cone $\tilde{V}$. The statement a) is proved. Let

$$\alpha = \begin{cases} x_n = 0, \\ z = 0 \end{cases}$$

be an (n-1)-dimensional plane described above. Denote by $\rho,\theta_1,\theta_2,\ldots,\theta_{n-2}$ the spherical coordinates in the plane α. Let S^{n-2} be the unit (n-2)-sphere in the plane α with the center at the point x_o. Clearly, for every point $y_o\in S^{n-2}$ there is a neighborhood $U(y_o)$ of a fixed size such that the spherical parameters (for example the geographical coordinates) can be chosen under the condition that the vectors

$$\vec{q},\ \frac{\partial\vec{q}}{\partial\theta},\ \ldots,\frac{\partial\vec{q}}{\partial\theta_{n-2}}$$

form an orthogonal frame in the (n-1)-plane at every point $z\in U(y_o)$, where $\vec{q}(\theta_1,\theta_2,\ldots,\theta_{n-2})$ is the positive vector-function of S^{n-2} with the origin at the point x_o.

Thus the convex cone $\tilde{V}$ is described locally by the vector-function

$$\vec{V}(\rho,\theta_1,\theta_2,\ldots,\theta_{n-2},s) = \overrightarrow{Q\bar{Q}}+s\vec{r}^{\,\tau}(\rho,\theta_1,\theta_2,\ldots,\theta_{n-2}),$$

where

$$\vec{r}(\rho,\theta_1,\theta_2,\ldots,\theta_{n-2}) = \rho\vec{q}(\theta_1,\theta_2,\ldots,\theta_{n-2})+ \qquad (*)$$
$$+b(0)\rho^{\tau+1}\vec{e}_n+[h_o(0)-h_{u_o}(0)-\delta\Delta h-b(0)\gamma^{-1}.\rho^{\tau+2}]\vec{e}_{n+1}$$

(*) Since x_o is the origin of Cartesian coordinates, we write $b(0)$ instead of $b(x_o)$.

and $0 \leqslant s \leqslant S_M$. Here we denote by M any point of the base of the convex cone V. Clearly $S_M \geqslant S_o$, where the number S_o depends only on δ and γ. We consider the description of S_o below. Since

$$\Delta h = h_o(0) - h_{u_o}(0),$$

then

$$\vec{V}(\rho, \theta_1, \theta_2, \ldots, \theta_{n-2}, s) =$$

$$= s[\rho\vec{q}(\theta_1, \theta_2, \ldots, \theta_{n-2}) + b(0)\rho^{\tau+2}\vec{e}_n + \tag{3.16}$$

$$+\{(1-\delta)\Delta h - \frac{b(0)}{\gamma}\rho^{\tau+2}\}\vec{e}_{n+1}] + 0\vec{\bar{Q}}$$

From (3.16) it follows that

$$\frac{\partial V}{\partial \rho} = s\vec{q} + sb(0)(\tau+2)\rho^{\tau+1}\vec{e}_n - \frac{sb(0)}{\gamma}(\tau+2)\rho^{\tau+1}\vec{e}_{n+1},$$

$$\frac{\partial \vec{V}}{\partial \theta_1} = s\rho\frac{\partial \vec{q}}{\partial \theta_1}, \frac{\partial \vec{V}}{\partial \theta_2} = s\rho\frac{\partial \vec{q}}{\partial \theta_2}, \ldots, \frac{\partial \vec{V}}{\partial \theta_{n-2}} = s\rho\frac{\partial \vec{q}}{\partial \theta_{n-2}}$$

$$\frac{\partial \vec{V}}{\partial s} = \vec{r} = \rho\vec{q} + b(0)\rho^{\tau+2}\vec{e}_n + [(1-\delta)\Delta h - \frac{b(0)}{\gamma}\rho^{\tau+2}]\vec{e}_{n+1}.$$

The vectors $\vec{q}, \frac{\partial \vec{q}}{\partial \theta_1}, \ldots, \frac{\partial \vec{q}}{\partial \theta_{n-2}}, \vec{e}_n, \vec{e}_{n+1}$ form an orthogonal frame in E^{n+1}. Now we find the normal $\vec{m}$ of $\tilde{V}$ at any point of this cone except at its vertex. The normals $\vec{m}$ used by the construction of the normal mapping have the component -1 at the vector $\vec{e}_{n+1}$, i.e.

$$\vec{m} = p_o\vec{q} + p_1\frac{\partial \vec{q}}{\partial \theta_1} + p_2\frac{\partial \vec{q}}{\partial \theta_2} + \ldots + p_{n-2}\frac{\partial \vec{q}}{\partial \theta_{n-2}} + \mu\vec{e}_n - \vec{e}_{n+1}. \tag{3.17}$$

This condition yields the equality

$$(s\rho)^{n-1}\rho^{\tau+1}b(0)(\tau+1) = 1 \tag{3.18}$$

at every point of the convex cone $\tilde{V}$ except at its vertex. Elementary calculations show that

$$p_1 = p_2 = \ldots = p_{n-2} = 0$$

at any point of $\tilde{V}$ except again its vertex. Since $|\vec{q}| = |\vec{e}_n| = 1$, then the normal image of $\tilde{V}$ contains the (n-1)-dimensional ball U^{n-1} with the center $(0,0,0,\dots,0,-|\mu|)$ and the radius $|p_o|$.

Omitting elementary calculations we obtain

$$|p_o| = \frac{\tau+2}{\tau+1} \frac{(1-\delta)\Delta h}{\rho} \tag{3.19}$$

$$|p_{n-1}| = \frac{1}{\rho^{\tau+2} b(0)(\tau+1)} \left|(1-\delta)\Delta h + \frac{b(0)}{\gamma}(\tau+1)\rho^{\tau+2}\right|. \tag{3.20}$$

The equalities (3.19) and (3.20) hold for any point of the convex cone $\tilde{V}$ except for its vertex. But $|p_o|$ and $|\mu|$ depend on the choice of the point of the cone $\tilde{V}$ where we consider the normal $\vec{m}$.

Now we choose our point $M \in \tilde{V}$ under the condition $M \in \beta' \cap \beta''$. Then from (3.3) and (3.4) it follows that

$$b(0)\rho^{\tau+2} = \gamma\Delta h.$$

Thus the radius of the ball U^{n-1} associated with the point $M \in \beta' \cap \beta''$ is as follows

$$|p_o| = \frac{\tau+2}{\tau+1} \frac{(1-\delta)(\Delta h)^{\frac{\tau+1}{\tau+2}} (b(0))^{\frac{1}{\tau+2}}}{\gamma^{\frac{1}{\tau+2}}} \tag{3.22}$$

and the center of this ball is the point $(0,0,0,\dots,-|\mu|)$, where

$$\mu = \frac{\tau+2-\delta}{(\tau+1)}$$

Lemma 1 is proved.

Note that

$$C' = \frac{\tau+2}{\tau+1}(1-\delta) \quad (\Delta h)^{\frac{\tau+1}{\tau+2}} [b(0)]^{\frac{1}{\tau+2}} \quad ;$$

$$C'' = \frac{\tau+2-\delta}{\tau+1} \quad . \qquad \square$$

Now we return to the proof of Theorem 1. We are concerned with the estimate of $\omega(R,V)$ from below. Clearly

$$\omega(R,V) \geqslant \omega(R,\widetilde{V}) \geqslant \int_{H^n} R(p)\,dp \geqslant C_o \int_{H^n} \frac{dp}{|p|^{2k}} ,$$

where H^n is the convex cone in $\widetilde{R}^n$ introduced in lemma 1.

Let us replace the convex cone H^n by the convex cone $\widetilde{H}^n$ symmetric to H^n with respect to the hyperplane $p_n = 0$. Then

$$\int_{H^n} \frac{dp}{|p|^{2k}} = \int_{\widetilde{H}^n} \frac{dp}{|p|^{2k}} .$$

Thus

$$\omega(R,V) \geqslant \omega(R,\widetilde{V}) \geqslant C_o\, \sigma_{n-2} \int_{\delta/\gamma}^{C''/\gamma} dp_n \int_0^{\psi(p_n,\gamma,\delta,\tau)} \frac{h^{n-2}dh}{(h^2+p^2)^k}$$

where

$$h^2 = p_1^2+p_2^2+ \dots +p_{n-1}^2 , \tag{3.23}$$

$$\psi(p_n,\gamma,\delta,\tau) = (p_n-\delta/\gamma)\frac{C'\gamma^{1-1/\tau+2}}{C''-\delta} \tag{3.24}$$

and σ_{n-2} is the area of the (n-2)-dimensional unit sphere. Note that $h \geqslant 0$ and $p_n \geqslant \delta/\gamma > 0$. Therefore

$$h^2+p_n^2 \leqslant (h+p_n)^2$$

and we obtain

$$\omega(R,V) \geqslant \omega(R,\widetilde{V}) \geqslant C_o\, \sigma_{n-2} \int_{\delta/\gamma}^{C''/\gamma} dp_n \int_0^{\psi(p_n,\gamma,\delta,\tau)} \frac{h^{n-2}dh}{(h+p_n)^{2k}} . \tag{3.25}$$

Let us introduce new variables

$$p = p_n\gamma, \quad q = h\gamma^{1/\tau+2} ;$$

then the inequalities (3.25) become

$$\omega(R,V) \geqslant \omega(R,\widetilde{V}) \geqslant$$

$$\geqslant C_o\, \sigma_{n-2}\, \gamma^{-1+2k-\frac{n-1}{\tau+2}} \int_\delta^{C''} dp \int_0^{\frac{C'(p-\delta)}{C''-\delta}} \frac{q^{n-2}dq}{\left(p+\gamma^{\frac{\tau+1}{\tau+2}}q^2\right)^{2k}} \tag{3.26}$$

Thus $\omega(R,V)$ is estimated from below. The estimate (3.26) completes part 2 of the proof of Theorem 1.

3.3 (Part 3). This part is devoted to the proof of the first statement of Theorem 1. First of all let us simplify the right part of the inequality (3.26) and combine both estimates (3.12) and (3.26) together. Since

$$0 \leqslant q \leqslant \frac{C'(p-\delta)}{C''-\delta},$$

then

$$\omega(R,V) \geqslant \omega(R,\tilde{V}) \geqslant C_o \, \sigma_{n-2} \, \gamma^{2k - \frac{n+\tau+1}{\tau+2}} \left(\frac{C'}{C''-\delta}\right)^{n-1} B_k(\delta,\gamma), \tag{3.27}$$

where

$$B_k(\delta,\gamma) = \int_{\delta}^{C''} \frac{(p-\delta)^{n-1}\,dp}{(p+\gamma)^{\frac{\tau+1}{\tau+2}} \left(\frac{C'(p-\delta)}{C''-\delta}\right)^{2k}}.$$

Since

$$\lim_{\delta\to 0} C' = \frac{\tau+2}{\tau+1} (\Delta h)^{\frac{\tau+1}{\tau+2}} [\, b(0)]^{\frac{1}{\tau+2}},$$

$$\lim_{\delta\to 0} C'' = \frac{\tau+2}{\tau+1},$$

and since we consider only sufficiently saml1 values of $\delta > 0$, then from (3.12) and (3.21) it follows that

$$d_2 \, \gamma^{\lambda + \frac{n+\tau+1}{\tau+2}} \geqslant \omega(R,V) \geqslant d_3 \gamma^{2k - \frac{n+\tau+1}{\tau+2}} B_k(\delta,\gamma)$$

where d_2 and d_3 are some positive constant independent of γ and δ.

The last inequality leads to the following one

$$1 \geqslant d_4 \gamma^{2k-\lambda-\frac{2(n+\tau+1)}{\tau+2}} B_k(\delta,\gamma), \tag{3.28}$$

where

$$d_4 = \frac{d_3}{d_2} = \text{const} > 0.$$

Thus the proof of the first statement of Theorem 1 is reduced to the corresponding estimates of $B_k(\delta,\gamma)$ from below for sufficiently small $\delta > 0$

and $\gamma > 0$.

First of all we simplify the expression for $B_k(\delta,\gamma)$ and then consider a few cases depending on values of the parameter k.

Let $t = p-\delta$ be the new variable of integration. Then the expression for $B_k(\delta,\gamma)$ becomes

$$B_k(\delta,\gamma) = \int_0^{C''-\delta} \frac{t^{n-1}\,dt}{(\mu t+\delta)^{2k}} \tag{3.29}$$

where

$$\mu = 1 + \frac{C'\gamma^{\frac{\tau+1}{\tau+2}}}{C''-\delta}$$

According to Assumption 2, the parameter k takes only non-negative values. Let us specially consider the case $k = 0$. Then

$$B_o(\delta,\gamma) = \int_0^{C''-\delta} t^{n-1}\,dt = \frac{(C''-\delta)^n}{n} \tag{3.30}$$

Let $\delta > 0$ be a fixed positive small number. It is sufficient to take $\delta = 1/2$, then $C''-\delta \geqslant 3/4$. Hence

$$B_o(1/2,\gamma) \geqslant (3/4)^n.1/n. \tag{3.31}$$

If $\delta < 1/2$, the number in the right part of (3.31) is only increasing. Thus the inequality (3.28) becomes

$$1 \geqslant d_4\gamma^{-\lambda-\frac{2(n+\tau+1)}{\tau+2}}(3/4)^n.1/n \tag{3.32}$$

for the case $k = 0$. Since γ can be chosen here as any small positive number, then (3.32) does not take place for any $\lambda \geqslant 0$ and $\tau \geqslant 0$. Thus the first statement of Theorem 1 is proved for the case $k = 0$.

Now let $k > 0$, then from the Hölder inequality it follows that

$$(\mu t+\gamma)^{2k} \leqslant (t^n+\delta^n)^{2k/n}(\mu^s+1)^{2k/s}$$

where

$1/n+1/s = 1$.

We obtain the following estimate for $B_k(\delta,\gamma)$ from below

$$B_k(\delta,\gamma) \geqslant \begin{cases} \dfrac{[\{(C''-\delta)^n+\delta^n\}^{1-2k/n} - \delta^{n(1-2k/n)}]}{n(\mu^s+1)^{2k/s}(1-2k/n)} & \text{if } 0<2k/n<1; \\[2ex] \dfrac{\ell n[1+C''-\delta)^n\delta^{-n}]}{n(\mu^s+1)^{2k/s}} & \text{if } 2k/n = 1; \\[2ex] \dfrac{\delta^{-n(2k/n-1)} - \{(C''-\delta)^n+\delta^n\}^{-2k/n-1)}}{n(\mu^s+1)^{2k/s}} & \text{if } 2k/n>1. \end{cases} \tag{3.33}$$

Let $0<\theta<1/6$ be some small number. We shall choose θ more precisely in the interval $(0,1/6)$. Now we set

$$\delta = \theta\,\frac{\tau+2}{\tau+1+\theta}\,. \tag{3.34}$$

Since

$$1 \leqslant \frac{\tau+2}{\tau+1+\theta} \leqslant 2$$

for all $\tau \geqslant 0$ and $0<\theta<1/6$, then

$$\theta < \delta < 2\theta\,. \tag{3.35}$$

From (3.15), it follows that

$$\delta = \theta . C''.$$

If we substitute the expression of δ in (3.33), then we obtain the following estimates for $\widetilde{B}_k(\theta,\gamma) = B_k(C''.\theta,\gamma) \equiv B_k(\delta,\gamma)$ from below

$$\widetilde{B}_k(\theta,\gamma) = \begin{cases} \dfrac{(C''^{n-2k}[\{(1-\theta)^n]^{1-2k/n} - \theta^{n(1-2k/n)}\}}{n(\mu^s+1)^{2k/s}(1-2k/n)} & \text{if } 0<2k/n<1; \\[2ex] \dfrac{\ell n[1+(\frac{1-\theta}{\theta})n]}{n(\mu^s+1)^{2k/s}} & \text{if } 2k/n = 1 \\[2ex] \dfrac{(C'')^{n-2k}}{n(\mu^s+1)^{2k/s}}\left[\dfrac{1}{\theta^{2k-n}} - \dfrac{1}{[(1-\theta)^n+\theta^n]^{2k-n/n}}\right] & \text{if } 2k/n>1 \end{cases} \tag{3.36}$$

The first part of Theorem 1 will be proved if we show that the inequality (3.28) implies a contradiction with the condition of this theorem. We establish it separately for the cases : a) $0<k<1$; b) $1\leqslant k<n/2$; c) $k = n/2$; d) $k>n/2$.

a) Assume that $0<k<1$. Then

$$2k - \lambda - \frac{2(n+\tau+1)}{\tau+2} < 0$$

for all $\lambda\geqslant 0$, $\tau\geqslant 0$ and $n\geqslant 2$. The inequality (3.35) and the equalities (3.36), (3.30), (3.14), (3.15) provide that

$$\lim_{\theta\to 0} \widetilde{B}_k(\theta,\gamma)\geqslant \frac{(\frac{\tau+2}{\tau+1})^{n-2k}}{(n-2k)[\,1+(1+\psi\gamma^{\frac{\tau+1}{\tau+2}})^s\,]^{2k/s}}$$

where

$$0<\psi = (\Delta h)^{\frac{\tau+2}{\tau+1}}(b(0))^{\frac{1}{\tau+2}}$$

(The definition of Δh and $b(0)$ were given by (3.2) and (1.5)).

The small parameter $\gamma>0$ is not definite up to now. Let $\gamma_o>0$ be a number such that

$$\psi\,\gamma_o^{\frac{\tau+1}{\tau+2}} = 1.$$

Then

$$1 + (1+\psi\gamma^{\frac{\tau+1}{\tau+2}})^s < 1+2^s$$

for all $0<\gamma\leqslant\gamma_o$. Recall that $s = n/n-1$. Thus

$$\lim_{\theta\to 0} \widetilde{B}_k(\theta,\gamma) \geqslant \frac{1}{(n-2k)(1+2s)^{2k/s}}\,(\frac{\tau+2}{\tau+1})^{n-2k} \tag{3.37}$$

for all $0<\gamma\leqslant\gamma_o$. Now let $\theta_o\in(0,1/6)$ be such a number that the inequality

$$\widetilde{B}_k(\theta_o,\gamma)\geqslant \frac{1}{2(n-2k)(1+2^s)^{2k/s}}\,(\frac{\tau+2}{\tau+1})^{n-2k} \tag{3.38}$$

holds for all $0<\theta\leqslant\theta_o$ and $0<\gamma\leqslant\gamma_o$. Thus the inequality (3.28) yields the following inequality

$$1 \geqslant d_4 \frac{1}{2(n-2k) 1+2^s)^{2k/s}} (\frac{\tau+2}{\tau+1})^{n-2k} \gamma^{2k-\lambda-\frac{2(n+\tau+1)}{\tau+2}} \tag{3.39}$$

for all $\theta \in (0,\theta_o]$ and $\gamma \in (0,\gamma_o]$.

But the inequality (3.39) does not hold for sufficiently small $\gamma \in (0,\gamma_o]$, because $n-2k>0$ and $k-\lambda/2-\frac{n+\tau+1}{\tau+2}<0$.* Thus the statement 1 of Theorem 1 is proved in the case a).

Now we introduce the number

$$K = \frac{n+\tau+1}{\tau+2} + \frac{\lambda}{2} . \tag{3.40}$$

The number K is very important in cases b), c), d).

b) <u>Assume that</u> $1 \leqslant k < n/2$. Then the following statement will be proved below : the inequality

$$k < K$$

provides the correctness of the statement 1 of Theorem 1, where n is the dimension of the convex domain G.

The case b) can be reduced to the final inequality (3.39) by the same considerations as in case a). Since the conditions $k<n/2$ and $k<K$ provide the inequalities

$$n - 2k > 0$$

and

$$k - \lambda/2 - \frac{n+\tau+1}{\tau+2} < 0$$

then the inequality (3.39) does not hold for sufficiently small $\gamma \in (0,\gamma_o]$ and $\theta \in (0,\theta_o]$.

Thus the statement 1 of Theorem 1 is also established in the case b).

---

* Recall that $k-\lambda/2-\frac{n+\tau+1}{\tau+2}<0$ for all $\lambda \geqslant 0$, $\tau \geqslant 0$ and $n \geqslant 2$, if $0<k<1$ (see the assumption of Theorem 1, §2).

c) Assume that $k = n/2$. Then the following statement will be proved in this section : The inequality

$$k \leqslant K$$

provides the correctness of the first statement of Theorem 1.

First consider the estimate for $\widetilde{B}_k(\theta,\gamma)$ from below for sufficiently small positive numbers θ and γ. Since $k = n/2$ and $1/n + 1/s = 1$, then

$$\widetilde{B}_k(\theta,\gamma) = \widetilde{B}_{n/2}(\theta,\gamma) = \frac{\ln\left[1 + \frac{(1-\theta)^n}{\theta^n}\right]}{n\left\{\left[1+\gamma^{\frac{\tau+1}{\tau+2}} \frac{C'}{C''(1-\theta)}\right]^{n/n-1} + 1\right\}^{n-1}} =$$

$$= \frac{\ln(\theta^n + (1-\theta)^n) - \ln\theta}{n\left[\left[1+\gamma^{\frac{\tau+1}{\tau+2}} \frac{(\tau-\delta+2)(1-\delta)(\Delta h)^{\frac{\tau+1}{\tau+2}}(b(0))^{\frac{1}{\tau+1}}}{(\tau+2)(1-\theta)}\right]^{n/n-1} + 1\right]^{n-1}}$$

Since

$$0 \leqslant \theta \leqslant 1/6$$

and

$$\theta \leqslant \delta \leqslant 2\theta,$$

then

$$1+\gamma^{\frac{\tau+1}{\tau+2}} \frac{(\tau+2-\delta)(1-\delta)(\Delta h)^{\frac{\tau+1}{\tau+2}} b(0)^{\frac{1}{\tau+1}}}{(\tau+2)(1-\tau)} \leqslant$$

$$\leqslant 1+\gamma^{\frac{\tau+1}{\tau+2}} (\Delta h)^{\frac{\tau+1}{\tau+2}} [b(0)]^{\frac{1}{\tau+1}}$$

Now let $\gamma_o > 0$ be a number such that

$$\gamma_o^{\frac{\tau+1}{\tau+2}} (\Delta h)^{\frac{\tau+1}{\tau+2}} [b(0)]^{\frac{1}{\tau+1}} = 1,$$

then the inequality

$$\tilde{B}_{n/2}(\theta,\gamma) \geq \frac{\ln(\theta^n+(1-\theta)^n)-n\ln\theta}{n[1+2^{n/n-1}]^{n-1}} \tag{3.41}$$

holds for all θ and γ satisfying the inequalities

$$0<\theta\leq 1/6,$$

$$0<\gamma\leq\gamma_o \ .$$

Thus the basic inequality (3.28) leads to the inequality

$$1\geq d_4\,\gamma^{2k-\lambda-\frac{2(n+\tau+1)}{\tau+2}}\ \frac{\ln[\theta^n+(1-\theta)^n]-n\ln\theta}{n[1+2^{n/n-1}]^{n-1}} \tag{3.42}$$

for $k = n/2$, $0<\theta\leq 1/6$, $0<\gamma\leq\gamma_o$. Now remember that the constant $d_4>0$ does not depend on θ and γ.

From the conditions of case c) it follows, that

$$k = n/2\leq K = \lambda/2 + \frac{n+\tau+1}{\tau+2} \ .$$

Hence

$$2k-\lambda-\frac{2(n+\tau+1)}{\tau+2}\leq 0. \tag{3.43}$$

Then we should consider two cases according to the inequality (3.43).

<u>The first one is</u>

$$2k-\lambda-\frac{2(n+\tau+1)}{\tau+2} < 0.$$

We set $\theta = 1/6$. Then the inequality (3.42) becomes

$$1\geq d_4\,\gamma^{2k-\lambda-\frac{2(n+\tau+1)}{\tau+2}}\ \frac{\ln(1+5^n)}{n[1+2^{n/n-1}]^{n-1}} \tag{3.44}$$

for all $\gamma\in(0,\gamma_o]$. But from (3.44) it follows that (3.44) does not hold if $\gamma>0$ is sufficiently small. Hence the statement c) is proved, if

$$2k-\lambda-\frac{2(n+\tau+1)}{\tau+2}<0.$$

<u>The second case is</u>

$$2k-\lambda-\frac{2(n+\tau+1)}{\tau+2}=0.$$

Then (3.42) takes the form

$$1\geqslant d_4\ \frac{\ell n[\theta^n+(1-\theta)^n]-n\ell n\theta}{n[1+2^{n/n-1}]^{n-1}} \tag{3.45}$$

if $0<\theta\leqslant 1/6$ and $0<\gamma\leqslant\gamma_o$. But

$$\lim_{\theta\to 0}\{[\ell n\theta^n+(1-\theta)^n]-n\ell n\theta\}=+\infty.$$

Therefore the inequality (3.45) is not valid if $\theta>0$ is sufficiently small. Thus the statement c) is proved in the second case. Hence the proof of the statement c) is completed.

d) <u>Finally assume that</u> $k>n/2$. Then the following statement holds : The inequality

$$k\leqslant K$$

provides the correctness of the statement 1 of Theorem 1. This statement can be proved in a similar way to case c). Therefore we omit this proof.

Thus part 3 of the proof of Theorem 1 is completed. We establish statement 2 of this theorem.

<u>3.4 (Part 4)</u>. This part is devoted to the proof of the second statement of Theorem 1 related to the uniform convergence of extended functions $U_m(x)$ to the extended function $U_o(x)$ (see statement of Theorem 1, section 2.3).

If our statement is not correct, then there exist the points $x_k\in G$ and positive integers $m_1<m_2<\ldots<m_k<\ldots$ such that

$$|U_o(x_k)-U_{m_k}(x_k)|\geqslant\varepsilon_o>0, \tag{3.46}$$

where $\varepsilon_o > 0$ is some number. Clearly the sequence of points $(x_k, U(x_k))$ is bounded in R^{n+1}.

Without losing generality we assume that the sequence of points $(x_k, U_{m_k}(x_k))$ converges to some point (x_o, U_o).

The point x_o belongs either to G or to ∂G. If x_o is an interior point of G, then we obtain

$$|U_o(x_o) - U_o| \geqslant \varepsilon_o \tag{3.47}$$

This inequality follows by passage of limit in the inequalities (3.46). But (3.47) cannot be valid because the point (x_o, U_o) has to lie on the graph of $U_o(x)$. Hence $U_o = U_o(x_o)$. This contradiction shows that $x_o \in \partial G$.

The conditions of Theorem 1 provide the application of the first statement of this theorem, proved in parts 1,2,3 and establish the following facts :

a) $h_o(x)$ is the border of the convex function $U_o(x)$;

b) $U_m(x)$ and $U_o(x)$ can be extended as continuous convex functions in the closed convex bounded domain $\bar{G} = G \cup \partial G$ and

$$U_m(x)\Big|_{\partial G} = h_{U_m}(x); \; U_o(x)\Big|_{\partial G} = h_o(x)$$

and the continuous functions $h_{U_m}(x)$ converge uniformly to the continuous function $h_o(x)$ on ∂G;

c) the convex functions $U_m(x)$ converge to the convex function $U_o(x)$ for all $x \in G \cup \partial G$. Hence

$$U_o = \lim U_{m_k}(x_k) = U_o(x_o) = h_o(x_o). \tag{3.48}$$

since $x_o \in \partial G$.

On the other hand if we take the limit in (3.46), then

$$|h_o(x_o) - U_o| \geqslant \varepsilon_o > 0. \tag{3.49}$$

But (3.48) and (3.49) are incompatible. This contradiction proves Theorem 1.

CHAPTER II. CONVEX GENERALIZED SOLUTIONS OF THE DIRICHLET PROBLEM FOR MONGE-AMPERE EQUATIONS.

In this chapter we are concerned with elliptic solutions of Monge-Ampère equations

$$\det(u_{ij}) = f(x,u,\nabla u).$$

These solutions can be either convex or concave functions. We shall consider only convex solutions of the Dirichlet problem. Existence theorems of generalized solutions of the Dirichlet problem for Monge-Ampère equations will be established here by significantly more general conditions than in our papers [2], [3], [4]. This can be achieved by means of the new, essentially stronger theorem of the uniform convergence of convex functions in a closed convex domain (see Theorem 1, §2,3, chapter 1).

§4. The Dirichlet problem for Monge-Ampère equations

$$\det(u_{ij}) = \frac{\varphi(x)}{R(\nabla u)} .$$

We consider this equation in a bounded open convex G and assume that $\varphi(x) \geqslant 0$ is summable in G and $R(p) > 0$ and locally summable in R^n.

4.1 The Dirichlet problem for weak convex solutions. The convex function $u(x) \in W^+(G)$ is called *a weak solution* of the equation

$$\det(u_{ij}) = \frac{\varphi(x)}{R(\nabla u)} \tag{4.1}$$

(see [1] - [4]) if

$$\omega(R,u,e) = \mu(e) \tag{4.2}$$

for every Borel subset e of G, where $\mu(e)$ is a given non-negative completely additive set function.

The equation (4.2) is the extension of the equation (4.1) to the class $W^+(G)$ of all convex functions u(x) defined in G. If

$$\mu(e) = \int_e \varphi(x)dx, \tag{4.3}$$

i.e. $\mu(e)$ is an absolutely continuous set function, then the weak solutions of (4.1.) have an absolutely continuous R-curvature and satisfy the equation (4.1) almost everywhere. Such weak solutions of equation (4.1) are called *generalized solutions* of (4.1). (see [2] - [4]).

In this subsection, we deal with weak solutions of the equation (4.1) and suppose that the following assumptions are fulfilled :

<u>Assumption 1</u>. G has a parabolic support of order not more than $\tau \geqslant 0$ (see §1, and also subsection 2.3).

<u>Assumption 2</u>. The function $R(p) > 0$ is locally summable in R^n and the inequality

$$R(p) \geqslant C_o |p|^{-2k} \tag{4.4}$$

holds for all $p \in R^n$ and $|p| \geqslant r_o$, where $k = \text{const} \geqslant 0$, $C_o = \text{const} > 0$, $r_o = \text{const} > 0$ (see subsection 2.3).

<u>Assumption 4</u>. Let x_o be an arbitrary point of ∂G, then there exists some n-ball $U_\rho(x_o)$ such that the inequality

$$\mu(e) \leqslant a[\sup_e(\text{dist}(x,\partial G))^\lambda]\text{mes } e \tag{4.5}$$

holds for every Borel set $e \subset U_\rho(x_o) \cap G$, where $\lambda = \text{const} \geqslant 0$ and $a = \text{const} > 0$.

<u>THEOREM 2</u>. Consider the Dirichlet problem

$$\omega(R,u,e) = \mu(e), \tag{4.6}$$

$$u\Big|_{\partial G} = h(x) \in C(\partial G). \tag{4.7}$$

Let the domain G and the functions R(p), $\mu(e)$ satisfy the assumptions 1,2,4. Let

$$K = \frac{n+\tau+1}{\tau+2} + \frac{\lambda}{2} \tag{4.8}$$

and let the following conditions

$$\begin{aligned} &k \leqslant K \quad \text{if} \quad k < 1 \quad \text{and} \quad k \geqslant n/2 \\ &k < K \quad \text{if} \quad 1 \leqslant k < n/2 \end{aligned} \tag{4.9}$$

and

$$\mu(G) < A(R)^{*} \tag{4.10}$$

be fulfilled too. Then the Dirichlet problem (4.6-7) has one and only one weak solution $u(x) \in W^{+}(G)$ and $h(x) = h_u(x)$, where $h_u(x)$ is the border of the function u(x).

PROOF. Denote by G_ε the open subdomain of G such that

$$\operatorname{dist}(x, G) > \varepsilon$$

for every $x \in G_\varepsilon$. The domain G_ε is not empty if $\varepsilon > 0$ is a sufficiently small number.

Now let $\mu(e)$ be a non-negative completely additive function of Borel subsets of G, satisfying the condition 4.10 :

$$\mu(G) < A(R).$$

We construct the family of set functions

$$\mu(e) = \mu(e \cap G_\varepsilon)$$

for all sufficiently small $\varepsilon > 0$. Clearly $\mu_\varepsilon(e)$ are non-negative completely additive functions of Borel subsets of G, satisfying the conditions

---

*According to (2.3) (see subsection 2.1) $A(R) = \int_{\tilde{R}^n} R(p)\,dp$.

$$\mu_\varepsilon(G) = \mu(G_\varepsilon) \leqslant \mu(G) < A(R), \tag{4.11}$$

$$\mu_\varepsilon(G \setminus G_\varepsilon) = 0. \tag{4.12}$$

The Dirichlet problem

$$\begin{cases} \omega(R, u_\varepsilon, e) = \mu_\varepsilon(e) \\ u_\varepsilon \Big|_{\partial G} = h(x) \end{cases} \tag{4.13}$$

has only one weak solution $u_\varepsilon(x) \in W^+(G)$ such that $h_{u_\varepsilon}(x) = h(x)$, where $h_{u_\varepsilon}(x)$ is the border of $u_\varepsilon(x)$. This fact is proved in [2] (see also [3]).* In the papers [2], [7] the comparison theorems and the estimates from which the following inequalities run out, are also proved :

$$u_{\varepsilon_2}(x) \leqslant u_{\varepsilon_1}(x), \tag{4.14}$$

if $0 < \varepsilon_2 \leqslant \varepsilon_1$, and

$$-B_o + \inf_{\partial G} h(x) \leqslant u_\varepsilon(x) \leqslant \sup_{\partial G} h(x) \tag{4.15}$$

for all admissible $\varepsilon > 0$** and for all $x \in G \cup \partial G$, where the constant B_o depends only on $\mu(G)$ or more precisely on the distinction $\mu(G)$ from $A(R)$.

From (4.14) and (4.15) it follows that the function u (x) converges pointwise to some convex function $u(x)$ for $x \in G$. Since all the functions $u_\varepsilon(x)$ have one and the same border $h_u(x) = h(x)$, then the conditions of Theorem 2 and Theorem 1 provide that $h(x)$ is the border of $u(x)$ and $u_\varepsilon(x)$ uniformly converge to $u(x)$ in $\bar{G}$. Finally $u(x)$ is the weak solution of the equation (4.6). This fact follows directly from the weak convergence of R-curvatures (see section 2.1). Theorem 2 is proved.

---

* The existence theorem for the Dirichlet problem (4.13)-(4.17) with the special functions $\mu_\varepsilon(G)$ is proved in [2], [3] only by assumption (4.10).

** $\varepsilon > 0$ is admissible if the domain G_ε is not empty.

4.2 The Dirichlet problem for generalized solutions. In this subsection, we consider the analogy of Theorem 2 for generalized solutions of the Dirichlet problem

$$\det(u_{ij}) = \frac{\varphi(x)}{R(\nabla u)} \tag{4.16}$$

$$u\Big|_{\partial G} = h(x) \in C(\partial G), \tag{4.17}$$

we only need to replace assumption 4 by the following simpler assumption 4'.

Assumption 4'. There exists some sufficiently small neighborhood U of G such that

$$0 \leqslant \varphi(x) \leqslant a(\text{dist}(x,\partial G))^{\lambda} \tag{4.18}$$

for all $x \in G \cap U$, where $a = \text{const} \geqslant 0$, $\lambda = \text{const} \geqslant 0$.

THEOREM 2'. Let all the conditions of Theorem 2 be fulfilled with only one distinction : assumption 4 is replaced by assumption 4', then the Dirichlet problem (4.16-17) has one and only one generalized solution $u(x) \in W^{+}(G)$ and h(x) is the border of the function u(x).

This theorem is the particular case of Theorem 2 and it does not need any special proof.

Recall that the case $\tau = 0$ corresponds to the case of a strictly convex domain. The existence theorems of the Dirichlet problem (4.16-4.17) (see [2], [3]) are particular cases of Theorem 2 and 2'.

§5. The Dirichlet problem for the Monge-Ampère equations $\det(U_{ij}) = f(x,u,\nabla u)$

This section is devoted to existence theorems for the Dirichlet problem

$$\det(u_{ij}) = f(x,u,\nabla u), \tag{5.1}$$

$$u\Big|_{\partial G} = h(x) \in C(\partial G). \tag{5.2}$$

We assume again that the problem (5.1-2) is considered in a bounded convex domain G in R^n, whose boundary ∂G satisfies the assumption 1 (see either subsection 2.3 or subsection 4.1). We shall investigate a few types of conditions for the function $f(x,u,p)$ providing the existence theorems of generalized solutions for the problem (5.1-2).

Assumption 5. $f(x,u,p)$ is continuous in $\bar{G} \times R \times R^n$ and the inequalities

$$0 \leqslant f(x,u,p) \leqslant \frac{\varphi(x)}{R(p)} \tag{5.3}$$

hold in the same domain $\bar{G} \times R \times \tilde{R}^n$, where $R(p)$ satisfies assumption 2 (see subsection 2.3) and $\varphi(x)$ is non-negative and summable in G and satisfies assumption 4' (see subsection 4.2).

Finally we formulate the concept that the generalized solution of the Dirichlet problem (5.1-2) has at least one generalized solution $u(x) \in W^+(G)$ if assumptions 1,2,4' and 5 are fulfilled and if the inequalities

$$\int_G \varphi(x)\,dx < A(R) \tag{5.4}$$

and either one of the following inequalities

$$k \leqslant K \quad \text{if } k < 1 \quad \text{and } k \geqslant n/2,$$

$$k < K \quad \text{if} \quad 1 \leqslant k < n/2$$

hold.

PROOF. We denote by $W_h^+(G)$ the set of all convex functions $u(x)$ satisfying the condition $h_u(x) = h(x)$, where $h_u(x)$ is the border of $u(x)$. The set $W_h^+(G)$ is not empty, because the Dirichlet problem

$$\det(z_{ij}) = \frac{\varphi(x)}{R(Dz)}$$

$$z\Big|_{\partial G} = h(x)$$

has the generalized solution $z(x) \in W_h^+(G)$. This follows directly from Theorem 2'.

Clearly, $W_h^+(G)$ is a convex set in the space $C(\bar{G})$. Let

$$F_u(x) = f(x,u,\nabla u)R(\nabla u) \tag{5.5}$$

for every function $u(x) \in W^+(G)$. The function $F_u(x)$ is non-negative in G and

$$F_u(x) \leqslant \varphi(x) \tag{5.6}$$

almost everywhere in G. If $u(x) \in W^+(G) \cap C^1(G)$, then $F_u(x) \in C(G)$ and $F_u(x) \in L(G)$.

If $u(x) \in W^+(G)$, then the same integral inequality can be proved by the approximation of functions $u_m(x) \in W^+(G) \cap C^1(G)$.

Hence the Dirichlet problem

$$\det(z_{ij}) = \frac{F_u(x)}{R(\nabla z)}, \tag{5.7}$$

$$z\Big|_{\partial G} = h(x) \tag{5.8}$$

has only one generalized solution $z(x) \in W_h^+(G)$, where $u(x) \in W_h^+(G)$.

This statement follows directly from Theorem 2', because all conditions of this theorem are valid. Thus we construct the operator

$$B : W_h^+(G) \to W_h^+(G)$$

and $z(x) = B(u(x))$. Clearly, the fixed points of operator B are generalized solutions of our initial Dirichlet problem (5.1-2).

The inequality

$$\int_G \varphi(x)\,dx < A(R)$$

yields the estimates

$$\inf_{\partial G} h(x) - B_o \leqslant z(x) = B(u(x)) \leqslant \sup_{\partial G} h(x)$$

where the constant B_o depends only on the distinction $\mu(G)$ from A(R) (see [3]). The set $B(W_h^+(G))$ is bounded in $C(\bar{G})$. Therefore we can take a sub-sequence

$$z_{i_m}(x) = B(u_{i_m}(x))$$

for each sequence $u_1(x), u_2(x), \ldots, u_m(x), \ldots \in W^+(G)$ converging for all $x \in G$. From the conditions of Theorem 1 it follows that the sequence $z_{i_m}(x)$ converges uniformly to some function $z_o(x) \in W_h^+(G)$. Thus the operator B is compact.

Let the functions $u_m(x) \in W_h^+(G)$ and let them converge to some function $u_o(x) \in W^+(G)$. Then the set of functions $v_m(x) = B(u_m(x))$ is compact in $W_h^+(G)$ considered as a subspace of $C(\bar{G})$. This fact is proved in just the same way as the compactness of the set $B(W_h^+(G))$.

We take some uniformly convergent sub-sequence $v_{n_k}(x)$ in $C(\bar{G})$. Let $\bar{v}(x)$ be the limit of this consequence. From the conditions of Theorem 4 and Theorem 1 it follows that $v(x) \in W_h^+(G)$.

Now using the properties of the convergent sequence of convex functions and their R-curvatures, we obtain that $v(x)$ and $v_o(x)$ are generalized solutions for one and the same equation

$$\det(v_{ij}) = \frac{F_{u_o}(x)}{R(Dv)} ,$$

and moreover the borders of $v(x)$ and $v_o(x)$ coincide. Therefore these functions coincide in G and operator B is continuous.

Now all the conditions of the Schauder principle are fulfilled and the initial Dirichlet problem (5.1-2) has at least one generalized solution. The theorem is proved.

The next theorem is the extension of Theorem 4 to wider classes of the non-negative functions $f(x,u,p)$ which can infinitely increase together with $|u| \to +\infty$.

This investigation is based on two assumptions. The first one is Assumption 1 (see subsection 2.3) requiring that the boundary ∂G of a bounded convex domain G has a parabolic support of order not more than $\tau = \text{const} \geqslant 0$. Some preliminary considerations are required for the statement of the second assumption. We denote by q_G the infimum of distancesbetween pairs of parallel supporting hyperplanes of G with the opposite outward normals. Let $\varepsilon > 0$ be any number less than $q_G/10$. We introduce the functions

$$\varphi(x,\lambda,\varepsilon) = \begin{cases} 1 & \text{if} \quad \operatorname{dist}(x,\partial G) \geqslant \varepsilon; \\ [\operatorname{dist}(x,\partial G)]^{\lambda} & \text{if } \operatorname{dist}(x,\partial G) < \varepsilon \end{cases} \tag{5.9}$$

and

$$R_k(p) = \begin{cases} 1 & \text{if} \quad |p| \leqslant 1; \\ \dfrac{1}{|p|^{2k}} & \text{if } |p| > 1 \end{cases} \tag{5.10}$$

where λ = const $\geqslant 0$ and $0 \leqslant k$ = const $< n/2$.

The statement of Assumption 6 is as follows : the function $f(x,u,p)$ is continuous in $\bar{G} \times R \times R^n$ and the inequalities

$$0 \leqslant f(x,u,p) \leqslant \frac{(a|u|+b)^{\alpha}}{R_k(p)} \varphi(x,\lambda,\varepsilon) \tag{5.11}$$

hold for all $x \in \bar{G}$, $u \leqslant 0$, $p \in \tilde{R}^n$, where a = const $\geqslant 0$, b = const $\geqslant 0$, $a^2+b^2 > 0$ and α = const 0.

THEOREM 3. Let Assumption 2 and 6 be fulfilled. Then the Dirichlet problem

$$\det(u_{ij}) = f(x,u,\nabla u) \tag{5.12}$$

$$u\Big|_{\partial G} = 0 \tag{5.13}$$

has at least one convex generalized solution if the inequalities

$$k < \frac{n+\tau+1}{\tau+2} + \frac{\tau}{2} \tag{5.14}$$

and

$$\alpha + 2k < n \tag{5.15}$$

hold.

PROOF. Let K be the set of all convex functions satisfying the boundary condition (5.13) in the classical meaning. Clearly, the set K is not empty and forms a convex cone in $C(\bar{G})$. All functions $u(x) \in K$ are non-positive in $\bar{G}$. Let

$$F_u(x) = f(x,u(x),Du(x)).R_k(Du(x)) \tag{5.16}$$

where $u(x) \in K$. From Assumption 6 it follows that $F_u(x)$ is non-negative and summable in G and

$$\int_G F_u(x)dx \leqslant \int_G (a|u(x)|+b)^\alpha dx \leqslant$$

$$\leqslant (a\|u\| + b)^\alpha \text{ mes } G, \tag{5.17}$$

where $\| u \| = \| u \|_{C(\bar{G})}$.

Now we consider the Dirichlet problem

$$\det(z_{ij}) = \frac{F_u(x)}{R_k(Dz)} \tag{5.18}$$

$$z\Big|_{\partial G} = 0. \tag{5.19}$$

Since $k < n/2$, then

$$A(R_k) = \int_{|p|\leqslant 1} dp + \int_{|p|>1} \frac{dp}{|p|^{2k}} = +\infty. \tag{5.20}$$

Therefore

$$\int_G F_u(x)dx \leqslant (a\,\|u(x)\| + b)^\alpha . \text{ mes } G < A(R_k). \tag{5.21}$$

The inequalities (5.21), (5.14), (5.15) together with Assumptions 1 and 6 provide the application of Theorem 2' to the Dirichlet problem (5.18-19). Thus we can say that this Dirichlet problem has only one generalized solution $z(x) \in K$. Hence the operator $A : K \to K$ arises such that

$$z(x) = A(u(x)), \tag{5.22}$$

where $z(x)$ is a generalized solution of the Dirichlet problem (5.18-19).

From theorem of estimates for convex function (see [3], [4]) it follows that

$$- T_{R_k}(\omega_u)\ \mathrm{diam}\ G \leqslant z(x) = A(u(x)) \leqslant 0 \qquad (5.23)$$

for all $x \in \bar{G}$, where

$$\omega_u = \int_G F_u(x)\,dx$$

and $T_{R_k}(\tau)$ is the inverse for the function

$$g_{R_k}(\rho) = \int_{|p| \leqslant \rho} R_k(p)\,dp =$$

$$= \mu_n(1 + \frac{n}{n-2k}(\rho^{n-2k} - 1)) \qquad (5.24)$$

(μ_n is the volume of the n-unit ball). Since

$$0 \leqslant \frac{2k}{n} < 1 \ , \ 0 < \frac{n-2k}{n} \leqslant 1$$

and

$$g_{R_k}(+\infty) = A(R_k) = +\infty,$$

we obtain

$$T_{R_k}(\tau) = [\,1 + \frac{n-2k}{n}\,(\frac{\tau}{\mu_n} - 1)]^{1/n-2k} \leqslant$$

$$\leqslant [\frac{\tau}{\mu_n} + 1]^{1/n-2k} \qquad (5.25)$$

for all τ $[0,+\infty)$.

Thus from (5.21 - 25) it follows that

$$\|A(u(x))\|_{C(\bar{G})} \leqslant \left[\frac{\omega_u + \mu_u}{\mu_n}\right]^{1/n-2k} .\ \mathrm{diam}\ G \leqslant$$

$$\leqslant [\frac{1}{\mu_n}(a\,\|u(x)\| + b)^{\alpha}\ \mathrm{mes}\ G + \mu_n]^{1/n-2k}\ \mathrm{diam}\ G. \qquad (5.26)$$

Therefore the operator $A : K \to K$ maps every bounded subset $Q \subset K$ in a bounded subset $A(Q)$ of K. Using the same considerations as in the proof of Theorem 4 we obtain that the operator $A : K \to K$ is compact and continuous.

Since $\alpha < n-2k$, $a \geqslant 0$, $b \geqslant 0$, $a^2 + b^2 > 0$, then one can find a positive sufficiently big number r_o such that the inequality

$$[\frac{1}{\mu_n}(a\,\|u(x)\|+b)^{\alpha}\ \text{mes}\ G+\mu_n]^{1/n-2k}\ \text{diam}\ G<r_o \tag{5.27}$$

holds.

Then from(5.26) it follows that

$$\|A(u(x))\|<\|u(x)\|$$

for all $u(x)\in S_{r_o}$, where S_{r_o} is the intersection of the cone K with the sphere $\|x\| = r_o$ in $C(\bar{G})$.

The compact and continuous operator A acts in the convex cone K and $A(Q)\subset Q$, where the convex set Q is the intersection of K with the closed ball $\|x\|\leqslant r_o$ in $C(\bar{G})$. Hence from the Schauder principle it follows that the Dirichlet problem (5.12-13) has at least one generalized solution $Z(x)\in C(\bar{G})$ and $\|Z(x)\|\leqslant r_o$. Theorem 3 is proved.

For other types of existence theorems of generalized and weak solutions of the Dirichlet problem for different classes of Monge-Ampère equations, see papers [3], [4].

References

[1] I. Bakelman, Generalized solutions of the Monge-Ampère equation, Dokl. Akad. Nauk USSR, 111 (1957), 1143-1145.

[2] I. Bakelman, Geometric methods of solution of elliptic equations, Monograph, Nauka, Moscou, (1965), 1-340.

[3] I. Bakelman, The Dirichlet problem for the n-dimensional Monge-Ampère equations and related problems in the theory of quasilinear elliptic equations, Proc. of Seminar held in Firenze Sept.-Oct. 1980, Ist. Naz. di Alta Matem. , Roma, (1982), 1-78.

[4] I. Bakelman, Elliptic generalized solutions of the Dirichlet problem for n-dimensional Monge-Ampère equations. Proc. of AMS Summer Inst. of Functional Analysis, Berkeley California 1983, University of Chicago (to appear).

[5] S.Y. Cheng and S.T. Yau, On the regularity of the solution of the Monge-Ampère equations $\det(u_{ij}) = F(x,u)$, Comm. Pure Appl. Mathem. 29, (1976), 495-516.

[6] A. Alexandrov, Convex polyhedra, Monograph, Moscow-Leningrad, (1950) 1-428.

[7] I. Bakelman, A. Werner, B. Kantor, Introduction to global differential geometry, Monograph, Nauka, Moscow, (1973), 1-440.

[8] A. Pogorelov, Exterior geometry of convex surfaces, Moscow, Nauka, (1969), 1-759.

[9] A. Pogorelov, The Minkowski multidimensional problem, J. Wiley and Sons, New York (1978).

[10] I. Bakelman, R-curvature, estimates and the stability of solutions of the Dirichlet problem for elliptic equations, Journal of Diff. Equations, 43 : 1, (1982), 106-133.

[11] I. Bakelman, Hypersurfaces with given mean curvature and quasilinear elliptic equations with strong non-linearities. Math. Sbornik, 75 (1968), 604-638; Math. U.S.S.R. Sb., 4, (1968), 561-596.

[12] I. Bakelman, Geometric problems in quasilinear elliptic equations, Upekhi Math. Nauk, 25 : 3, (1970), 49-112; Russian Math. Surveys (1970), 25, N° 3, 45-109.

[13] J. Serrin, The Dirichlet problem for the minimal surface equation in higher dimensions, J. Reine Angew. Math. 223 (1968), 170-187.

[14] J. Serrin, The problem of Dirichlet for quasilinear elliptic differential equations with many independent variables. Philos. Trans. Roy. Soc., (1969), 413-496.

[15] S.N. Bernstein, Collection of papers, vol. 3, Akad. of Sc. U.S.S.R., (1960), 1-439.

[16] O.A. Ladyzhenskya, N.N. Uraltzeva, Linear and quasilinear elliptic equations, Nauka, (1973), 2nd ed., 1-576.

[17] L. Caffarelli, L. Nirenberg, J. Spruck, The Dirichlet problem for non-linear second order elliptic equations I. Monge-Ampère equations, Comm. Pure Appl. Math. (to appear).

[18] A. Alexandrov, Existence and uniqueness of convex surface with prescribed integral curvature, Dokl. Akad. of Sc. U.S.S.R. 35, 8, (1942).

[19] I. Bakelman, On the theory of Monge-Ampère equations, Vestnik LGU, N° 1, (1958), 25-38.

[20] A. Friedman, Variational principles and free-boundary problems, J. Wiley and Sons, (1982).

[21] S. Mikhlin, Variationals methods of solving linear and non-linear boundary value problems; Diff. equations and their Applications, Proc. of the Conference, Prague, (1962) Czechosl. Akad. of Sc.

Ilya J. Bakelman
Texas A & M University
College Station
Texas 11843
U.S.A.
and
Institute for Advanced Study
Princeton
New Jersey 08540
U.S.A.

L BARTHÉLEMY & P BÉNILAN

Phénomène de couche limite pour la pénalisation d'équations quasi-variationnelles elliptiques avec conditions de Dirichlet au bord

Considérons l'inéquation quasi-variationelle (I.Q.V.)

$$\left.\begin{array}{l} v \in H_0^1(\Omega),\ v \leqslant Mv \\ \text{et pour tous } w \in H_o^1(\Omega) \text{ avec } w \leqslant Mv \\ \int \text{grad } v.\ \text{grad}(w-v) \geqslant \int (f-v)(w-v) \end{array}\right\} \quad (1)$$

où Ω est un ouvert borné de $\mathbb{R}^N$, $f \in L^\infty(\Omega)$ et pour tout $v : \Omega \to \mathbb{R}$,

$$Mv(x) = 1 + \inf_{\substack{y \in \Omega \\ y \geqslant x}} \text{ess } v(y)$$

Une méthode classique (cf. [3]) pour étudier (1) consiste à l'approcher par le problème "pénalisé"

$$\left.\begin{array}{l} u_\varepsilon \in H_0^1(\Omega) \cap L^\infty(\Omega) \\ u_\varepsilon - \Delta u_\varepsilon + \frac{1}{\varepsilon}(u_\varepsilon - Mu_\varepsilon)^+ = f \quad \text{dans } \mathcal{D}'(\Omega) \end{array}\right\} \quad (2)$$

Ce problème admet pour tout $f \in L^\infty(\Omega)$ une solution unique et la"suite" $\{u_\varepsilon\}_{\varepsilon > 0}$ est décroissante lorsque $\varepsilon \downarrow 0$, bornée dans $L^\infty(\Omega)$; elle converge donc dans $L^p(\Omega)$ pour tout $1 \leqslant p < \infty$: notons $u(f)$ sa limite, qui est définie pour tout $f \in L^\infty(\Omega)$.

Lorsque (1) admet une solution, $u(f)$ est l'unique solution de (1); il est même prouvé dans [1] que (1) admet une solution si et seulement si $u(f) \in H_o^1(\Omega)$ ou encore si et seulement si $\{u_\varepsilon\}$ est bornée dans $H^1(\Omega)$. Ceci est toujours réalisé lorsque $f \geqslant 0$; mais en général pour f de signe quelconque, (1) n'admet pas de solution, $\{u_\varepsilon\}$ n'est pas bornée dans $H^1(\Omega)$ et plus précisément la condition au bord , $u_\varepsilon = 0$ sur $\partial\Omega$, disparaît lorsque $\varepsilon \to 0$: il y a phénomène de couche limite.

Pour se convaincre de ce phénomène de couche limite, le lecteur pourra vérifier par un calcul élémentaire la situation dans le cas $\Omega =]0,1[\subset \mathbb{R}$ et $f \equiv -k$ où $k \in \mathbb{R}$: il existe en effet une constante $k_o > 0$ * telle que (1) admette une solution si et seulement si $k \leqslant k_o$; et lorsque $k > k_o$, la limite $u = u(f)$ qui appartient à $C^\infty([0,1])$ n'est pas nulle en 0 : on a $u(0) = 1 + \min u < 0$.

Nous nous proposons dans cet article de caractériser pour $f \in L^\infty(\Omega)$ quelconque, la limite $u(f)$ de $\{u_\varepsilon\}_{\varepsilon > 0}$.
Nous montrons que $u(f)$ est solution maximum du problème

$$\left.\begin{array}{l} u - \Delta u \leqslant f,\ u \leqslant Mu \quad \text{sur } \Omega \\ u \leqslant 0 \text{ sur } \partial\Omega \end{array}\right\} \tag{3}$$

en un sens que nous préciserons. Nous montrerons aussi, en supposant Ω suffisamment régulier, que u est l'unique solution de

$$\left.\begin{array}{l} u \in H^1_{loc}(\Omega) \cap C(\bar{\Omega}) \\ u - \Delta u \leqslant f \quad \text{dans } D'(\Omega) \\ u \leqslant Mu \text{ sur } \Omega \text{ et } u - \Delta u = f \text{ dans } D'(\{u < Mu\}) \\ u = Mu \wedge 0 \text{ sur } \partial\Omega \end{array}\right\} \tag{4}$$

Notons que la même situation se rencontrera si l'on considère une inéquation variationnelle (I.V.)

$$\left.\begin{array}{l} v \in H^1_o(\Omega),\ v \leqslant \psi \\ \text{et pour tout } w \in H^1_o(\Omega) \text{ avec } w \leqslant \psi \\ \int \text{grad } v.\ \text{grad}(w-v) \geqslant \int (f-v)(w-v) \end{array}\right\} \tag{5}$$

lorsque l'obstacle ψ n'est pas positif ou nul sur $\partial\Omega$.
L'I.V. (5) peut apparaître comme un cas particulier de (1) en posant pour

---

* $k_o = \dfrac{e+1}{(\sqrt{e}-1)^2}$

tout $v : \Omega \to R$, $Mv = \psi$.
Nous obtiendrons les mêmes résultats dans ce cas.

En effet nous utiliserons principalement des techniques de T-accrétivité dans $L^\infty(\Omega)$ ou $C(\bar{\Omega})$ permettant de traiter le problème (1) avec une application M de $L^\infty(\Omega)$ dans $L^\infty(\Omega)$ beaucoup moins particulière que celle considérée ci-dessus. Nous remplacerons également l'opérateur $I - \Delta$ par un opérateur différentiel d'ordre 2 à coefficients discontinus : sous cette forme générale, les résultats obtenus nous semblent nouveaux même dans le cas $f \geq 0$.

Dans la section I nous précisons le cadre de notre travail et énonçons les résultats; les sections II et III sont consacrées aux démonstrations.

Les techniques utilisées permettent de traiter d'autres problèmes analogues : par exemple, on peut remplacer la condition de Dirichlet par une condition mixte au bord, on peut aussi considérer un opérateur différentiel non linéaire; enfin on peut étudier le problème parabolique soit par pénalisation, soit par la théorie des semi-groupes dans L^∞. Plus généralement on peut développer un cadre abstrait regroupant tous ces résultats qui fera l'objet d'un article ultérieur.

I. NOTATIONS, HYPOTHESES et RESULTATS

Nous nous donnons Ω un ouvert borné de $\mathbb{R}^N$ de bord $\partial\Omega$ et un opérateur différentiel du second ordre

$$A = -\sum_{i,j} \frac{\partial}{\partial x_i} \left(a_{ij} \frac{\partial}{\partial x_j}\right) + \sum_i a_i \frac{\partial}{\partial x_i} + a_o$$

où les coefficients sont des fonctions mesurables sur Ω vérifiant

$$(H1) \begin{cases} a_{ij} = a_{ji} \in L^\infty(\Omega) \text{ et il existe } \alpha > 0 \text{ tel que} \\ \sum_{i,j} a_{ij}(x)\xi_i\xi_j \geq \alpha \sum_i \xi_i^2 \quad \forall \xi \in \mathbb{R}^N, \text{ p.p.} x \in \Omega \\ a_o \in L^1(\Omega),\ \beta = \inf_\Omega \text{ess } a_o > 0 \text{ et } \gamma = \sup_\Omega \text{ess } a_o^{-1} \sum a_i^2 < \infty \end{cases}$$

Il résulte des hypothèses que pour $u \in H^1_{loc}(\Omega) \cap L^\infty_{loc}(\Omega)$, on peut définir la distribution Au qui appartient à $H^{-1}_{loc}(\Omega) + L^1_{loc}(\Omega)$.

Etant donnés $u, v \in H^1_{loc}(\Omega) \cap L^\infty_{loc}(\Omega)$, nous noterons

$$A(u,v) = \sum_{i,j} a_{ij} \frac{\partial u}{\partial x_j} \frac{\partial v}{\partial x_i} + \sum_i a_i \frac{\partial u}{\partial x_i} v + a_o uv$$

qui est une fonction de $L^1_{loc}(\Omega)$ et lorsque $A(u,v) \in L^1(\Omega)$, nous poserons

$$a(u,v) = \int A(u,v)dx$$

En particulier $a(u,v)$ est défini pour $u, v \in H^1(\Omega) \cap L^\infty(\Omega)$ et pour $u,v \in H^1_{loc}(\Omega) \cap L^\infty_{loc}(\Omega)$ avec $\text{supp } u \cap \text{supp } v$ compact dans Ω.
Nous noterons A la trace de $\mathcal{A}$ sur $(H^1_o(\Omega) \cap L^\infty(\Omega)) \times L^\infty(\Omega)$, c'est-à-dire que A est l'opérateur de $L^\infty(\Omega)$ défini par $Au = \mathcal{A}u$ sur

$$D(A) = \{u \in H^1_o(\Omega) \cap L^\infty(\Omega) ; \mathcal{A}u \in L^\infty(\Omega)\} \tag{6}$$

Nous utiliserons les hypothèses complémentaires

(H2) (a) $a_o \in L^\nu(\Omega)$ avec $\nu > \frac{N}{2}$

(b) Ω vérifie la condition du cône extérieur *

Sous l'hypothèse (H2.a), d'après les inclusions de Sobolev, la forme bilinéaire $a(u,v)$ est continue sur $H^1_o(\Omega) \times H^1_o(\Omega)$ et il existe $\lambda_o \in \mathbb{R}$ tel que

$$a(u,u) + \lambda_o \|u\|^2_{L^2} \geq \frac{\alpha}{2} \|u\|_{H^1} \quad \text{pour tout } u \in H^1_o(\Omega) \tag{7}$$

D'après le théorème de De Giorgi-Nash (voir [4]), sous les hypothèses (H2) pour $f_o \in L^\nu(\Omega), f_1, \ldots, f_N \in L^{2\nu}(\Omega)$, toute solution de

$$u \in H^1_o(\Omega), \ \mathcal{A}u = f_o - \sum_i \frac{\partial f_i}{\partial x_i} \quad \text{dans } \mathcal{D}'(\Omega)$$

est dans $C_o(\Omega) = \{u \in C(\bar{\Omega}) ; u = 0 \text{ sur } \partial\Omega\}$. En particulier sous les

---

* c'est-à-dire que pour tout $z \in \partial\Omega$, il existe un cône ouvert non vide C de $\mathbb{R}^N$ et $r > 0$ tel que $(z + C) \cap B(z,r) \cap \Omega = \phi$.

hypothèses (H2)

$$D(A) = \{u \in C_o(\Omega) \cap H_o^1(\Omega)\ ;\ Au \in L^\infty(\Omega)\}$$

Etant donné $f \in L^\infty(\Omega)$, on appellera *sous-solution de*

$$E_\infty(f) \qquad Au = f \text{ sur } \Omega,\ u = 0 \text{ sur } \partial\Omega$$

toute fonction u satisfaisant

$$I_\infty(f) \begin{cases} u \in H^1_{loc}(\Omega) \cap L^\infty(\Omega) \\ Au \leqslant f \quad \text{dans } \mathcal{D}'(\Omega) \\ (u-v)^+ \in H_o^1(\Omega) \text{ pour tout } v \in D(A) \end{cases}$$

La dernière condition dans $I_\infty(f)$ est une expression faible de la condition au bord : " $u \leqslant 0$ sur $\partial\Omega$". Nous montrerons (voir lemme 4) que sous les hypothèses (H2), $I_\infty(f)$ est impliquée par

$$I'_\infty(f) \begin{cases} u \in H^1_{loc}(\Omega) \cap L^\infty(\Omega) \\ Au \leqslant f \quad \text{dans } \mathcal{D}'(\Omega) \\ \limsup_{x \to z} \text{ess } u(x) \leqslant 0 \text{ pour tout } z \in \partial\Omega \end{cases}$$

Nous nous donnons également une application

$$M : L^\infty(\Omega) \to L^\infty(\Omega)$$

vérifiant les trois conditions suivantes

$$(M1) \begin{cases} \text{(a) M est une T-contradiction de } L^\infty(\Omega), \text{ c'est-à-dire} \\ \|(Mu-M\hat{u})^+\|_\infty \leqslant \|(u-\hat{u})^+\|_\infty \text{ pour tout } u,\hat{u} \in L^\infty(\Omega)\ ^* \\ \text{(b) pour toute suite décroissante bornée } \{u_n\} \text{ de } L^\infty(\Omega) \\ \text{convergeant p.p. vers u, on a } Mu = \lim Mu_n \text{ p.p. sur } \Omega \\ \text{(c) il existe } \ell_o,\ f_o \in L^\infty(\Omega) \text{ tels que } \ell_o \text{ soit sous-solution de} \\ E_\infty(f_o) \text{ et } \ell_o \leqslant M\ell_o \text{ p.p. sur } \Omega \end{cases}$$

---

* Pour tout $1 \leqslant p \leqslant \infty$, $\|.\|_p$ désigne la norme dans l'espace de Lebesque $L^p(\Omega)$; pour toute fonction u, $u^+ = \max(u,0)$

L'exemple le plus classique est l'application M donnée par

$$Mu(x) = k(x) + \inf_{\substack{y\in\mathbb{R}^{+N} \\ x+y\in\Omega}} \text{ess}\, (u(x+y)-h(y)) \tag{8}$$

avec $k\in L^{\infty}(\Omega)$ et $h\in L^{\infty}(\mathbb{R}^{+N})$ vérifiant inf ess $k \geqslant$ sup ess h.

Notons aussi que pour tout $\psi\in L^{\infty}(\Omega)$, l'application M définie par $Mu = \psi$ pour tout $u\in L^{\infty}(\Omega)$, satisfait (M1); la propriété (c) est vérifiée par

$$\ell_o = - \|\psi\|_{\infty}, f_o = 0.$$

Enonçons d'abord le résultat d'existence et d'unicité du problème pénalisé :

PROPOSITION 1. Sous les hypothèses (H1) et (M1.a), pour tout $f\in L^{\infty}(\Omega)$ et tout $\varepsilon>0$, il existe une unique solution $u_{\varepsilon}(f)$ du problème

$$E_{\varepsilon}(f,M) \qquad u_{\varepsilon}\in D(A),\ Au_{\varepsilon} + \frac{1}{\varepsilon}(u_{\varepsilon}-Mu_{\varepsilon})^{+} = f$$

De plus pour f, $\hat{f}\in L^{\infty}(\Omega)$

$$\|(u_{\varepsilon}(f) - u_{\varepsilon}(\hat{f}))^{+}\|_{\infty} \leqslant \frac{1}{\beta}\|(f-\hat{f})^{+}\|_{\infty} \tag{9}$$

et si on suppose aussi (M1.b.c) alors

$$u_{\varepsilon}(f) \geqslant \ell_o - \frac{1}{\beta}\ \|(f_o - f)^{+}\|_{\infty} \tag{10}$$

D'après (9), il est facile de voir que

$$\varepsilon_1 < \varepsilon_2 \Rightarrow u_{\varepsilon_1}(f) \leqslant u_{\varepsilon_2}(f) \text{ p.p. sur } \Omega \tag{11}$$

(noter que $u_{\varepsilon_2}(f) = u_{\varepsilon_1}(g)$ avec $g = f + (\frac{1}{\varepsilon_1} - \frac{1}{\varepsilon_2})(u_{\varepsilon_2}(f) - Mu_{\varepsilon_2}(f))^{+})$

On peut ainsi affirmer, en tenant compte de (10), que sous les hypothèses

(M1), il existe $u(f) \in L^\infty(\Omega)$ tel que $u_\varepsilon(f) \downarrow u(f)$ lorsque $\varepsilon \downarrow 0$.* Enonçons alors les

THEOREME 1. Sous les hypothèses (H1) et (M1), pour tout $f \in L^\infty(\Omega)$ il existe une solution maximum du problème

$$I(f,M) \quad \begin{cases} u \in H^1_{loc}(\Omega) \cap L^\infty(\Omega) \\ Au \leqslant f \quad \text{dans } \mathcal{D}'(\Omega) \\ u \leqslant Mu \text{ p.p. sur } \Omega \\ (u - v)^+ \in H^1_o(\Omega) \text{ pour tout } v \in D(A) \end{cases}$$

De plus cette solution maximum est $u(f) = \downarrow \lim u_\varepsilon(f)$ et est aussi la solution maximum du problème

$$IV(f,M) \quad \begin{cases} u \in L^\infty(\Omega),\ u \in Ku \quad \text{et} \\ a(u, \zeta^2(w-u)) \geqslant \int f\, \zeta^2(w-u)dx \quad \forall \zeta \in \mathcal{D}(\Omega),\ \forall w \in K(u) \\ (u-v)^+ \in H^1_o(\Omega) \text{ pour tout } v \in D(A) \end{cases}$$

où pour tout $u \in L^\infty(\Omega)$

$$K(u) = \{w \in H^1_{loc}(\Omega) \cap L^\infty_{loc}(\Omega);\ w \leqslant Mu \text{ p.p. sur } \Omega\} \tag{12}$$

THEOREME 1'. Supposant (H2) en plus des hypothèses (H1) et (M1), pour tout $f \in L^\infty(\Omega)$, la solution maximum de $I(f,M)$ est aussi la solution maximum de

$$I'(f,M) \quad \begin{cases} u \in H^1_{loc}(\Omega) \cap L^\infty(\Omega) \\ Au \leqslant f \quad \text{dans } \mathcal{D}'(\Omega) \\ u \leqslant Mu \text{ p.p. sur } \Omega \\ \limsup\operatorname{ess}_{x \to z} u(x) \leqslant 0 \quad \text{pour tout } z \in \partial\Omega \end{cases}$$

---

* La notation $u_\varepsilon \downarrow u$ lorsque $\varepsilon \downarrow 0$, en abrégé $u = \downarrow \lim u_\varepsilon$, signifie que pour toute suite (ε_n) décroissant vers 0, la suite $\{u_{\varepsilon_n}\}$ décroît et tend vers u p.p. sur Ω

et encore la solution maximum de

$$\text{IV}'(f,M)\quad \begin{cases} u \in L^\infty(\Omega),\ u \in K(u) \text{ et} \\ a(u,\zeta^2(w-u)) \geqslant \int f\,\zeta^2(w-u) \quad \forall\zeta \in \mathcal{D}(\Omega),\ \forall w \in K(u) \\ \limsup\limits_{x\to z} \text{ess } u(x) \leqslant 0 \quad \text{pour tout } z \in \partial\Omega \end{cases}$$

Introduisons maintenant les hypothèses complémentaires sur l'application M :

$$(\text{M2})\quad \begin{cases} \text{(a) } M \text{ envoie } C(\bar\Omega) \text{ dans lui-même} \\ \text{(b) } M \text{ est concave sur } C(\bar\Omega), \text{ c'est-à-dire} \\ \lambda Mu + (1-\lambda)Mv \leqslant M(\lambda u + (1-\lambda)v) \quad \forall u,v \in C(\bar\Omega),\ \lambda \in [0,1] \\ \text{(c) il existe } c_o \in \mathbb{R} \text{ telle que pour toute constante } c \leqslant c_o, \\ \quad Mc(x) > c \quad \text{pour tout } x \in \bar\Omega \end{cases}$$

Notons que l'application M définie par (8) vérifie ces hypothèses sous réserve que

$$k \in C(\bar\Omega),\ h \in C(\mathbb{R}^{+N}),\ \min_{\bar\Omega} k > \max_{\bar\Omega - \bar\Omega \cap \mathbb{R}^{+N}} h \tag{13}$$

et que Ω soit"suffisamment convexe"(voir [3]).
D'autre part pour tout $\psi \in C(\bar\Omega)$ il est clair que l'application $M : u \in L^\infty(\Omega) \to \psi \in C(\bar\Omega)$ vérifie toujours (M2).
Enonçons le

THEOREME 2. Sous les hypothèses (H1), (H2), (M1) et (M2), pour tout $f \in L^\infty(\Omega)$, la solution maximum de I(f,M) est l'unique solution du problème

$$\text{E}(f,M)\quad \begin{cases} u \in C(\bar\Omega) \cap H^1_{loc}(\Omega) \\ Au \leqslant f \quad \text{dans } \mathcal{D}'(\Omega) \\ u \leqslant Mu \quad \text{sur } \bar\Omega \\ Au = f \quad \text{dans } \mathcal{D}'(\{x \in \Omega; u(x) < Mu(x)\}) \\ u = Mu \wedge 0 \text{ sur } \partial\Omega \end{cases}$$

REMARQUE. Comme nous l'ont fait remarquer L. Tartar, la condition $\sup_\Omega \text{ess}\ a_o^{-1} \Sigma a_1^2 < \infty$ de (H1) peut être remplacée par $\Sigma a_i^2 \in L^1(\Omega)$ sous réserve de l'unicité de la solution de

$$u \in H_o^1(\Omega) \cap L^\infty(\Omega),\ u \geq 0,\ Au = 0 \text{ dans } \mathcal{D}'(\Omega).$$

Cette unicité est assurée si $\Sigma a_i^2 \in L^\nu(\Omega)$ avec $\nu > \frac{N}{2}$ et Ω vérifie (H2,b); par contre nous ignorons la situation dans le cas général.

II. DEMONSTRATIONS RELATIVES AUX THEOREMES 1.

Nous reprenons pour l'essentiel le schéma de la première preuve de Bensoussan-Lions pour l'étude des I.Q.V. développé dans un cadre abstrait dans [1] et [2]. La différence réside d'une part dans la généralité des hypothèses (H1) et d'autre part dans la nécessité de travailler dans l'espace $H^1_{loc}(\Omega)$ au lieu de $H^1(\Omega)$.
L'utilisation de la T-accréticité dans $L^\infty(\Omega)$ permettra de contourner cette difficulté.

Montrons d'abord

LEMME 1. Sous les hypothèses (H1), soient $f \in L^\infty(\Omega)$ et $u \in H^1_{loc}(\Omega) \cap L^\infty(\Omega)$ tels que

$$Au \leq f \quad \text{dans } \mathcal{D}'(\Omega) \quad \text{et } u^+ \in H_o^1(\Omega) \tag{14}$$

Alors

$$\|u^+\|_\infty \leq \frac{1}{\beta} \|f^+ \chi_{\{u>0\}}\|_\infty \tag{15}$$

$$\|\text{grad } u^+\|_2 \leq c\ \|f^+ \chi_{\{u>0\}}\|_\infty \tag{16}$$

et pour tout ouvert ω relativement compact dans Ω,

$$\|\text{grad } u\|_{L^2(\omega)} \leq c(\omega)(\|f^+\|_\infty + \|u^-\|_\infty) \tag{17}$$

où la constante c (resp. $c(\omega)$) ne dépend que de $\alpha,\gamma,\beta,\|a_o\|_1$, $|\Omega|$ (resp. et $\|a_{ij}\|_\infty$, $\text{dist}(\omega,\partial\Omega)$).

PREUVE. Notons d'abord qu'étant donnés $v \in H^1(\Omega) \cap L^\infty(\Omega)$ et $1 \leq p < \infty$, on a $v^{+p} \in H^1(\Omega) \cap L^\infty(\Omega)$ et pour tout $\theta > 0$,

$$a(v,v^{+p}) = \int (pv^{+p-1} \Sigma a_{ij} \frac{\partial v}{\partial x_i} \frac{\partial v}{\partial x_j} + v^{+p} \Sigma a_i \frac{\partial v}{\partial x_i} + a_o v^{+p+1})dx$$

$$\geq \int (pv^{+p-1} [\Sigma a_{ij} \frac{\partial v}{\partial x_i} \frac{\partial v}{\partial x_j} - \theta \Sigma (\frac{\partial v}{\partial x_i})^2] + v^{+p+1} [a_o - \frac{\Sigma a_i^2}{4p\theta}]) dx$$

d'ou d'après (H1),

$$a(v,v^{+p}) \geq p(\alpha-\theta) \int v^{+p-1} |\text{grad } v|^2 dx + (1 - \frac{\gamma}{4p\theta}) \int a_o v^{+p+1} dx \tag{18}$$

Considérons maintenant u vérifiant les hypothèses du lemme. D'après l'inégalité de Kato *

$$Au^+ \leq f\chi_{\{u>0\}} \leq f^+ \text{ dans } \mathcal{D}'(\Omega)$$

et donc pour tout $1 \leq p < \infty$, puisque $u^{+p} \in H_o^1(\Omega) \cap L^\infty(\Omega)$,

$$a(u^+, u^{+p}) \leq \int u^{+p} f^+ dx \leq \|u^+\|_{p+1}^p \|f^+ \chi_{\{u>0\}}\|_{p+1}$$

Appliquant (18) avec $v = u^+$, $\theta = \alpha$ on obtient pour tout $\frac{\gamma}{4\alpha} \leq p < \infty$, compte-tenu de $a_o \geq \beta$ p.p. sur Ω,

$$\beta(1 - \frac{\gamma}{4p\alpha}) \|u^+\|_{p+1}^{p+1} \leq \|u^+\|_{p+1}^p \|f^+ \chi_{\{u>0\}}\|_{p+1}$$

---

* Etant donnés $u \in H^1_{loc}(\Omega) \cap L^\infty_{loc}(\Omega)$ et $f \in L^1_{loc}(\Omega)$ tels que $Au \leq f$ dans $\mathcal{D}'(\Omega)$, on a l'inégalité de Kato, $Au^+ \leq f\chi_{\{u>0\}}$ dans $\mathcal{D}'(\Omega)$. En effet donnons-nous $\theta \in C^\infty(\mathbb{R})$ avec $\theta = 0$ sur $]-\infty,0]$, $\theta = 1$ sur $[1,\infty[$ et $\theta' \geq 0$ sur $\mathbb{R}$; étant donné $\zeta \in \mathcal{D}(\Omega)^+$ on a pour tout $n \in \mathbb{N}$

$\int f\zeta\theta(nu)\,dx \geq a(u,\zeta\theta(nu)) \geq \int \theta(nu)A(u,\zeta)\,dx$, d'où à la limite

$\int f\chi_{\{u>0\}}\zeta\,dx \geq \int \chi_{\{u>0\}}A(u,\zeta)\,dx = a(u^+,\zeta)$.

d'où (15) en divisant par $\|u^+\|^p_{p+1}$ et faisant $p\to\infty$.

Appliquant (18) avec $v = u^+$, $p = 1$ et $\theta = \frac{\alpha}{2}$, on obtient

$$\frac{\alpha}{2}\|\text{grad } u^+\|_2^2 \leqslant \int u^+ f^+ \, dx + \left(\frac{\gamma}{2\alpha} - 1\right)\int a_o u^{+2} \, dx$$

d'ou (16) avec $c^2 = \frac{2}{\alpha\beta}\left(|\Omega| + \left(\frac{\gamma}{2\alpha} - 1\right)^+ \frac{\|a_o\|_1}{\beta}\right)$.

Pour prouver (17), posons $m = \|u^-\|_\infty$ et fixons $\zeta \in \mathcal{D}(\Omega)$ tel que $0 \leqslant \zeta \leqslant 1$ et $\zeta = 1$ sur ω. On a

$$\int f^+(u+m)\,dx \geqslant \int f\,\zeta^2(u+m)\,dx \geqslant a(u, \zeta^2(u+m)) =$$

$$\int (\Sigma a_{ij}\frac{\partial u}{\partial x_j}[\zeta^2\frac{\partial u}{\partial x_i} + 2\zeta(u+m)\frac{\partial\zeta}{\partial x_i}] + [\Sigma a_i \frac{\partial u}{\partial x_i} + a_o u]\,\zeta^2(u+m))\,dx$$

$$\geqslant \int [\frac{\alpha}{2}|\zeta \text{ grad } u|^2 - \frac{2}{\alpha}(u+m)^2(\sum_j(\sum_i a_{ij}\frac{\partial\zeta}{\partial x_i})^2 + 4\,\zeta^2\Sigma a_i^2)$$

$$+ a_o\,\zeta^2 u(u+m)]\,dx \geqslant \frac{\alpha}{2}\|\text{grad } u\|^2_{L^2(\omega)} - \frac{2}{\alpha}\|u+m\|^2_\infty\,(4\gamma\,\|a_o\|_1$$

$$+ \|\text{grad }\zeta\|_2^2 \max_{i,j}\|a_{ij}\|_\infty) - m^2\,\|a_o\|_1$$

d'ou (17) compte-tenu de $\|u+m\|_\infty \leqslant \frac{1}{\beta}\|f^+\|_\infty + m$; la constante $c(\omega)$ dépend de $\alpha, \beta, \gamma, \|a_{ij}\|_\infty, \|a_o\|_1$ et de

$$\text{cap }\bar{\omega} = \inf\{\|\text{grad }\zeta\|_2 ;\ \zeta \in \mathcal{D}(\Omega),\ 0 \leqslant \zeta \leqslant 1,\ \zeta = 1 \text{ sur } \bar{\omega}\}$$

que l'on peut évaluer à l'aide de $\text{dist}(\omega, \partial\Omega)$. ∎

REMARQUE 1. Comme corollaire immédiat du lemme 1, pour toute suite $\{u_n\}$ de sous-solutions de $E_\infty(f)$ avec $\sup_n \|u_n^-\|_\infty < \infty$, on a

$\{u_n\}$ bornée dans $H^1_{loc}(\Omega) \cap L^\infty(\Omega)$

$\{(u_n - v)^+\}$ bornée dans $H^1_0(\Omega)$ pour tout $v \in D(A)$

et donc toute valeur d'adhérence u (pour la topologie $*$-faible) de $L^\infty(\Omega)$ est encore sous-solution de $E_\infty(f)$. ∎

Rappelons maintenant qu'un opérateur C de $L^\infty(\Omega)$ est dit T-*accrétif* si

$$\|(u-\hat{u})^+\|_\infty \leqslant \|(u-\hat{u}+\lambda(Cu-C\hat{u}))^+\|_\infty \quad \forall u,\ \hat{u}\in D(C),\ \forall\lambda>0.$$

Il est m-T-*accrétif* si de plus

$$R(1+\lambda C) \overset{\text{déf}}{=} \{u+\lambda Cu;\ u\in D(C)\} = L^\infty(\Omega)$$

pour tout $\lambda>0$; il suffit d'ailleurs que ce soit vrai pour au mons un $\lambda_o>0$. Sous les hypothèses (M1.a), pour $\varepsilon>0$ l'opérateur de pénalisation

$$B_\varepsilon\ :\ u\in L^\infty(\Omega) \to \frac{1}{\varepsilon}(u-Mu)^+\in L^\infty(\Omega)$$

est T-accrétif et lipschitzien (voir [2]). La première partie de la proposition 1 résulte alors immédiatement de la propriété :
la somme d'un opérateur m-T-accrétif et d'un opérateur T-accrétif lipschitzien est m-T-accrétif (voir [1]) et du

<u>LEMME 2</u>. Sous les hypothèses (H1), l'opérateur $\tilde{A} = A-\beta 1$ est m-T-accrétif dans $L^\infty(\Omega)$.

<u>PREUVE</u>. Etant donné $\lambda>0$, appliquons le lemme 1 à l'opérateur $A^\lambda = \lambda A+(1-\lambda\beta)$ dont le coefficient $a_o^\lambda = a_o+(1-\lambda\beta)$ vérifie $a_o^\lambda\geqslant 1$ p.p. sur Ω; on aura d'après (15),

$$\|u^+\|_\infty \leqslant \|(u+\lambda(Au-\beta u))^+\|_\infty \text{ pour tout } u\in D(A)$$

d'où la T-accrétivité de l'opérateur linéaire $\tilde{A} = A-\beta 1$.

Sous l'hypothèse complémentaire (H2.a), la m-T-accrétivité de $\tilde{A}$ est assurée puisqu'alors A est continu de $H_o^1(\Omega)$ dans $H^{-1}(\Omega)$ et pour λ_o suffisamment grand $A+\lambda_o$ est coercif. Dans le cas général approchons A par les opérateurs A^n dont les coefficients sont

$$a_{ij}^n = a_{ij},\ a_o^n = \min(a_o,n),\ a_i^n = a_i\,\chi_{\{a_o\leqslant n\}}$$

Ils vérifient les mêmes hypothèses (H1) et $a_o^n\in L^\infty(\Omega)$; pour $f\in L^\infty(\Omega)$, il existe $u_n \in H_o^1(\Omega) \cap L^\infty(\Omega)$ solution de $A^n u_n = f$ dans $\mathcal{D}'(\Omega)$, qui par utilisation du lemme 1 vérifie

$$\|u_n\|_\infty \leq \frac{1}{\beta} \|f\|_\infty , \ \|\text{grad } u_n\|_2 \leq c \ \|f\|_\infty ;$$

considérant alors u valeur d'adhérence de $\{u_n\}$, on a

$$u \in H_o^1(\Omega) \cap L^\infty(\Omega) \text{ et } Au = f \text{ dans } \mathcal{D}'(\Omega);$$

donc $\tilde{A}$ est m-T-accrétif. ■

La deuxième partie de la proposition 1 résulte alors du

LEMME 3. Sous les hypothèses (H1) et (M1), pour tout $f \in L^\infty(\Omega)$ et toute sous-solution v de $E_\infty(f)$ telle que $v \leq Mv$ p.p. sur Ω, on a

$$v \leq u_\varepsilon(f) \text{ p.p. sur } \Omega \text{ pour tout } \varepsilon > 0.$$

Il suffit en effet alors pour prouver (10) d'écrire

$$u_\varepsilon(f) \geq u_\varepsilon(f_o) - \frac{1}{\beta} \|(f_o - f)^+\|_\infty \geq \ell_o - \frac{1}{\beta} \|(f_o - f)^+\|_\infty .$$

PREUVE DU LEMME 3. Fixons $f \in L^\infty(\Omega)$ et $\varepsilon > 0$, et construisons par récurrence la suite $\{u_n\}$ définie par

$$Au_o = f, \ Au_n + \frac{1}{\varepsilon}(u_n - Mu_{n-1})^+ = f$$

L'existence et l'unicité d'une telle suite $\{u_n\}$ et le fait qu'elle est décroissante résulte de la m-T-accrétivité de $\tilde{A} = A - \beta I$ et de l'hypothèse (M1.a) : if suffit en fait d'appliquer la première partie de la proposition 1 en remplaçant M par l'application constante $u \to Mu_{n-1}$.

Soit alors v une sous-solution de $E_\infty(f)$ telle que $v \leq Mv$.
Alors $A(v - u_n) \leq g_n$ dans $\mathcal{D}'(\Omega)$ où

$$g_o = 0 \text{ et } g_n = \frac{1}{\varepsilon} (u_n - Mu_{n-1})^+ \text{ pour } n \geq 1$$

On a aussi par hypothèse $(v - u_n)^+ \in H_o^1(\Omega)$ pour tout $n \geq 0$. D'après le lemme 1, on a donc $\|(v-u_n)^+\|_\infty \leq \frac{1}{\beta}\|g_n\, \chi_{\{v>u_n\}}\|_\infty$.

En particulier $v \leq u_o$; d'autre part

$$v \leq u_{n-1} \Rightarrow v \leq Mv \leq Mu_{n-1} \Rightarrow (u_n - Mu_{n-1})^+ \leq (u_n - v)^+$$

$$\Rightarrow g_n\, \chi_{\{v>u_n\}} = 0 \Rightarrow v \leq u_n.$$

Donc par récurrence $v \leq u_n$ pour tout n.

If suffit alors de montrer que $u_n \downarrow u_\varepsilon(f)$. Notant $\bar{u} = \downarrow \lim u_n$, on a d'après (M1.b), $M\bar{u} = \downarrow \lim Mu_{n-1}$ et donc

$$Au_n \to f - \frac{1}{\varepsilon}(\bar{u} - M\bar{u})^+ \quad \text{p.p. sur } \Omega$$

avec

$$\|Au_n\|_\infty \leq c = \|f\|_\infty + \frac{1}{\varepsilon}\|(u_o - M\bar{u})^+\|_\infty.$$

D'après (16) du lemme 1, $\bar{u} \in H_o^1(\Omega)$ et $u_n \to \bar{u}$ dans $H_o^1(\Omega)$; donc $\bar{u} = u_\varepsilon(f)$. ■

PREUVE DU THEOREME 1. Fixons $f \in L^\infty(\Omega)$ et notons $u_\varepsilon = u_\varepsilon(f)$. Comme nous l'avons noté dans la section 1, la'suite'$\{u_\varepsilon\}$ décroît lorsque ε décroît, et d'après (10), $u_\varepsilon \geq -m = -(\|\ell_o^-\|_\infty + \frac{1}{\beta}\|(f_o - f)^+\|_\infty)$.
On peut donc définir $u = \downarrow \lim u_\varepsilon$. D'après le lemme 2, pour toute sous-solution v de $E_\infty(f)$ avec $v \leq Mv$, on a $u_\varepsilon \geq v$ et donc à la limite $u \geq v$. Enfin il est clair que toute solution de IV(f,M) est solution de I(f,M) de telle sorte qu'il suffit de prouver que u est bien solution de IV(f,M).

Puisque $Au_\varepsilon \leq f$ dans $\mathcal{D}'(\Omega)$, d'après (17) du lemme 1, on a $u \in H^1_{loc}(\Omega)$ et $u_\varepsilon \to u$ dans $H^1_{loc}(\Omega)$. En particulier $Au_\varepsilon \to Au$ dans $\mathcal{D}'(\Omega)$ et donc

$$(u_\varepsilon - Mu_\varepsilon)^+ = \varepsilon(f - Au_\varepsilon) \to 0 \text{ dans } \mathcal{D}'(\Omega);$$

d'après (M1.b), $Mu = \downarrow \lim Mu_\varepsilon$ et donc

$$(u_\varepsilon - Mu_\varepsilon)^+ \to (u - Mu)^+ \text{ dans } L^2(\Omega)$$

d'où l'on déduit $u \leq Mu$ p.p. sur Ω. En d'autres termes $u \in K(u)$.
Aussi étant donné $v \in D(A)$, on a $A(u_\varepsilon - v) \leq f - Av$ dans $\mathcal{D}'(\Omega)$; d'après (16)

du lemme 1, $(u-v)^+ \in H_0^1(\Omega)$. Enfin soit $w \in K(u)$ et $\zeta \in \mathcal{D}(\Omega)$; on a $w \leq Mu \leq Mu_\varepsilon$, d'où $(u_\varepsilon - Mu_\varepsilon)^+(w-u_\varepsilon) \leq 0$ et donc

$$a(u_\varepsilon, \zeta^2(w-u_\varepsilon)) \geq \int f\, \zeta^2(w-u_\varepsilon)\, dx$$

Passant à la limite

$$a(u,\zeta^2(w-u)) \geq \overline{\lim_\varepsilon}\, a(u_\varepsilon,\zeta^2(w-u_\varepsilon)) \geq \int f\zeta^2(w-u)\, dx. \quad \blacksquare$$

Sous l'hypothèse (H2), $u_\varepsilon \in C_0(\Omega)$ et donc avec les notations ci-dessus $\limsup_{x\to z} \operatorname{ess} u(x) \leq 0$ pour tout $z \in \partial\Omega$.

Le théorème 1' est alors un corollaire immédiat du théorème 1, grâce au

<u>LEMME 4</u>. Sous les hypothèses (H1) et (H2), pour tout $f \in L^\infty(\Omega)$, toute solution u de $I'_\infty(f)$ est solution de $I_\infty(f)$.

<u>PREUVE DU LEMME 4</u>. Considérant u solution de $I'_\infty(f)$, nous avons à montrer que $(u-v)^+ \in H_0^1(\Omega)$ pour tout $v \in D(A)$; remplaçant u par $u-v$ et f par $f-Av$, il suffit de prouver que $u^+ \in H_0^1(\Omega)$.

Etant donné $\delta > 0$, $(u-\delta)^+$ est à support compact et donc $(u-\delta)^+ \in H_0^1(\Omega)$; d'après l'inégalité de Kato, $A(u-\delta)^+ \leq (f-\delta a_0)\,\chi_{\{u>\delta\}} \leq f^+$. Appliquent (16) du lemme 1, puisque $(u-\delta)^+ \uparrow u^+$ lorsque $\delta \downarrow 0$, on a bien $u^+ \in H_0^1(\Omega)$.

III. PREUVE DU THEOREME 2

Nous supposons vérifiées les hypothèses (H1) et (H2) et nous fixons $f \in L^\infty(\Omega)$. Nous notons $\bar{u}$ la solution de

$$\bar{u} \in H_0^1(\Omega) \cap C_0(\Omega),\ A\bar{u} = f \text{ dans } \mathcal{D}'(\Omega)$$

Pour tout $\psi \in L^\infty(\Omega)$, l'application $u \in L^\infty(\Omega) \to \psi \in L^\infty(\Omega)$ vérifiant (M1), d'après le théorème 1', l'inéquation

$$
I'(f,\psi)\begin{cases} v\in H^1_{loc}(\Omega)\cap L^\infty(\Omega) \\ Av\leq f \quad \text{dans } \mathcal{D}'(\Omega) \\ v\leq\psi \text{ p.p. sur } \Omega \\ \limsup_{x\to z} \text{ess } v(x)\leq 0 \text{ pour tout } z\in\partial\Omega \end{cases}
$$

admet une solution maximum que nous notons $v(\psi)$. Utilisant la définition de $v(\psi)$, il est immédiat que

$$
\lambda_1 v(\psi_1)+\lambda_2 v(\psi_2)\leq v(\lambda_1\psi_1+\lambda_2\psi_2) \quad \forall\psi_1,\psi_2\in L^\infty(\Omega),\ \lambda_1,\lambda_2\geq 0,\ \lambda_1+\lambda_2\leq 1 \tag{19}
$$

$$
\psi_1\leq\psi_2+m\Rightarrow v(\psi_1)\leq v(\psi_2)+m \quad \forall\psi_1,\psi_2\in L^\infty(\Omega),\ m\geq 0 \tag{20}
$$

On a aussi

$$
v(\psi)\leq\bar{u} \text{ p.p. sur } \Omega \quad \forall\psi\in L^\infty(\Omega) \tag{21}
$$

Nous allons prouver le

LEMME 5. Sous les hypothèses (H1) et (H2), pour tout $\psi\in C(\bar{\Omega})$ on a $v = v(\psi)$ si et seulement si v est solution de

$$
E(f,\psi)\begin{cases} v\in C(\bar{\Omega})\cap H^1_{loc}(\Omega) \\ Av\leq f \text{ dans } \mathcal{D}'(\Omega) \\ v\leq\psi \quad \text{sur } \Omega \\ Av = f \text{ dans } \mathcal{D}'(\{x\in\Omega;\ v(x)<\psi(x)\}) \\ v = \psi\wedge 0 \text{ sur } \partial\Omega \end{cases}
$$

Notons que ce lemme est le cas particulier du théorème 2 pour l'application $u\in L^\infty(\Omega)\to\psi\in C(\bar{\Omega})$. Donnons d'abor la

PREUVE DU THEOREME 2. L'application M vérifiant (M1) et (M2) étant fixée, considérons $u(f)$ la solution maximum de $I'(f,M)$. Puisque $u = u(f)$ vérifie $I'(f,Mu)$ on a $u\leq v(Mu)$; alors $Mu\leq M(v(Mu))$ et, puisque $v(Mu)\leq Mu$, $v(Mu)$ est solution de $I'(f,M)$ d'où $u = v(Mu)$. En d'autres termes, $u = u(f)$ est point fixe de l'application $w \in L^\infty(\Omega) \to v(Mw) \in L^\infty(\Omega)$; nous noterons S cette application :

$$Sw = v(Mw) \text{ pour tout } w \in L^\infty(\Omega)$$

D'après l'hypothèse (M2.a) et le lemme 5, S applique $C(\bar{\Omega})$ dans $C(\bar{\Omega})$ et il est clair que u est solution de $E(f,M)$ si et seulement si u est point fixe de la restriction de S à $C(\bar{\Omega})$. Pour prouver le théorème 2, il suffit donc de montrer que S admet un unique point fixe qui appartient à $C(\bar{\Omega})$.
Notons que d'après (20), v et donc aussi S sont des T-contractions.

Considérons une constante (négative) $\underline{u}$ telle que

$$\underline{u} \leq c_o \wedge \left(-\frac{1}{\beta} \|f^-\|_\infty\right) \tag{22}$$

où c_o est la constante de (M2.c) et soit $w \in L^\infty(\Omega)$ telle que $w \geq \underline{u}$ p.p. sur Ω. Puisque $\underline{u} \leq c_o$, on a $M\underline{u} > \underline{u}$ et donc $Mw \geq \underline{u}$ p.p. sur Ω; puisque $\underline{u} \leq -\frac{1}{\beta}\|f^-\|_\infty$,

$$A\underline{u} = \underline{u}a_o \leq -\|f^-\|_\infty \leq f \text{ dans } \mathcal{D}'(\Omega)$$

et donc $v(\underline{u}) = \underline{u}$; utilisant (20), on en déduit

$$Sw = v(Mw) \geq v(\underline{u}) = \underline{u}$$

Utilisant de plus (21), on a ainsi montré que pour toute constante $\underline{u}$ satisfaisant (22), S applique $[\underline{u},\bar{u}] = \{w \in L^\infty(\Omega);\ \underline{u} \leq w \leq \bar{u}\}$ dans lui-même. Pour prouver le théorème 2, il suffit de montrer qu'étant donné $\underline{u}$ constante quelconque vérifiant (22), la restriction de S à $[\underline{u},\bar{u}]$ admet un unique point fixe qui appartient à $C(\bar{\Omega})$.

Fixons $\underline{u}$ constante vérifiant (22). D'après (M2.c), il existe $k \in [0,1[$ telle que

$$k\underline{u} + (1-k)\bar{u} \leq M\underline{u} \quad \text{sur } \Omega \tag{23}$$

(il suffit de prendre $k = \left(1 - \dfrac{\min M\underline{u} - \underline{u}}{\max \bar{u}^+ - \underline{u}}\right)^+$)

Puisque $k\underline{u} + (1-k)\bar{u}$ est sous-solution de $E_\infty(f)$,

$$k\underline{u} + (1-k)\bar{u} \leqslant S\underline{u} \quad \text{sur } \Omega$$

Notant maintenant que, d'après (M2.b) et (19), la restriction de S à $C(\bar{\Omega})$ est concave, on peut raisonner comme dans Hanouzet-Joly [5], déduire qu'étant donnés $w_1, w_2 \in [\underline{u},\bar{u}] \cap C(\bar{\Omega})$

$$k\alpha(\underline{u} - Sw_2) \leqslant Sw_1 - Sw_2 \leqslant k\beta(Sw_1 - \underline{u})$$

où $\alpha,\beta \in [0,1]$ satisfont

$$\alpha(\underline{u} - w_2) \leqslant w_1 - w_2 \leqslant \beta(w_1 - \underline{u})$$

pour montrer que la restriction de S à $[\underline{u},\bar{u}] \cap C(\bar{\Omega})$ admet un point fixe unique u qui est limite uniforme des suites $(S^n w)_n$ pour tout $w \in [\underline{u},\bar{u}] \cap C(\bar{\Omega})$. Considérant enfin $w \in [\underline{u},\bar{u}]$, on aura

$$S^n\underline{u} \leqslant S^n w \leqslant S^n\bar{u} \quad \text{p.p. sur } \Omega$$

et donc $(S^n w)_n$ converge aussi dans $L^\infty(\Omega)$ vers le point fixe u de S dans $[\underline{u},\bar{u}] \cap C(\bar{\Omega})$. ■

Pour la preuve du lemme 5 nous allons utiliser le

LEMME 6. Sous les hypothèses (H1) et (H2.a), l'ensemble

$$\{u \in C(\bar{\Omega}) \cap H^1(\Omega);\ Au \in L^\infty(\Omega)\}$$

est dense dans $C(\bar{\Omega})$.

Ce résultat de densité est évident lorsque $a_{ij} \in W^{1,\infty}(\Omega)$ et $a_i \in L^\infty(\Omega)$ puisqu'alors $Au \in L^\infty(\Omega)$ pour tout $u \in W^{2,\infty}(\Omega)$; n'ayant pas trouvé de référence dans le cas général où nous nous plaçons, nous en donnons une preuve plus loin.

PREUVE DU LEMME 5. Considérons d'abord v solution de $E(f,\psi)$. Il est clair que v vérifie $I'(f,\psi)$ et donc $v \leqslant v(\psi)$; puisque $v(\psi) \leqslant \psi$ p.p. sur Ω, il suffit de prouver que

$$v = v(\psi) \text{ p.p. sur } \omega = \{x \in \Omega;\ v(x) < \psi(x)\}. \tag{24}$$

Or $u = v(\psi) - v \in H^1_{loc}(\omega) \cap L^\infty(\omega)$, $Au \leqslant 0$ dans $\mathcal{D}'(\omega)$ et $\limsup_{x \to z} \text{ess } u(x) \leqslant 0$ pour tout $z \in \partial\omega$; donc, appliquant (15) du lemme 1 à $u - \delta$ pour tout $\delta > 0$, il en résulte que $u \leqslant 0$ p.p. sur ω, d'où (24).

Nous avons donc maintenant à montrer l'existence d'une solution de $E(f,\psi)$. Montrons d'abord que l'on peut se ramener à $f = 0$ p.p. sur Ω et $\psi \geqslant 0$ sur $\partial\Omega$. En effet étant donné $\psi \in C(\bar{\Omega})$, on peut résoudre (voir [4], théorème 8.30)

$$w \in H^1_{loc}(\Omega) \cap C(\bar{\Omega}),\ Aw = f \text{ dans } \mathcal{D}'(\Omega),\ w = \psi \wedge 0 \text{ sur } \partial\Omega.$$

Maintenant v est solution de $E(f,\psi)$ si et seulement si $\bar{v} = v - w$ est solution de $E(0,\bar{\psi})$ avec $\bar{\psi} = \psi - w$ qui vérifie $\bar{\psi} = \psi^+ \geqslant 0$ sur $\partial\Omega$.

Supposons donc $\psi \geqslant 0$ sur $\partial\Omega$. La résolution de $E(0,\psi)$ est alors classique au moins sous des hypothèses de régularité sur les coefficients de A et sur Ω. Pour être complet nous en reprenons rapidement les étapes.

Supposons d'abord $\psi \in H^1(\Omega)$ et $A\psi \geqslant -g$ dans $\mathcal{D}'(\Omega)$ avec $g \in L^\infty(\Omega)$. Notant v_ε la solution du problème pénalisé $Av_\varepsilon + \frac{1}{\varepsilon}(v_\varepsilon - \psi)^+ = 0$, on a

$$v_\varepsilon - \psi \in H^1(\Omega) \cap L^\infty(\Omega),\ (v_\varepsilon - \psi)^+ \in H^1_o(\Omega)$$

et

$$A(v_\varepsilon - \psi) + \frac{1}{\varepsilon}(v_\varepsilon - \psi)^+ \leqslant g \quad \text{dans } \mathcal{D}'(\Omega);$$

utilisant l'estimation (18) de la preuve du lemme 1 avec $\theta = \alpha$, on obtient pour $p \geqslant \frac{4\alpha}{\gamma}$,

$$\frac{1}{\varepsilon}\int (v_\varepsilon - \psi)^{+p+1} \leqslant \int g(v_\varepsilon - \psi)^{+p}$$

d'où il résulte à la limite lorsque $p \to \infty$,

$$\|Av_\varepsilon\|_\infty = \|\frac{1}{\varepsilon}(v_\varepsilon - \psi)^+\|_\infty \leqslant \|g\|_\infty$$

D'après le théorème de De Giorgi, on en déduit que $\{v_\varepsilon\}$ est relativement compacte dans $C_o(\Omega)$ et donc converge uniformément vers $v(\psi)$. Il est clair alors que $v = v(\psi)$ est solution de $E(0,\psi)$: $v \in C_o(\Omega)$ et $Av = 0$ dans

$\mathcal{D}'(\{v<\psi\})$ puisque, pour tout ouvert $\omega \subset\subset \{v<\psi\}$ on a pour ε suffisamment petit, $v_\varepsilon \leq \psi$ sur ω et donc $Av_\varepsilon = 0$ dans $\mathcal{D}'(\omega)$.

Soit maintenant $\psi \in C(\bar{\Omega})$ avec $\psi \geq 0$ sur $\partial\Omega$ quelconque. D'après le lemme 6, il existe $\psi_n \in C(\bar{\Omega}) \cap H^1(\Omega)$ avec $A\psi_n \in L^\infty(\Omega)$ tels que $\psi_n \to \psi$ uniformément sur $\bar{\Omega}$; remplaçant éventuellement ψ_n par $\psi_n + \|\psi_n - \psi\|_\infty$ on peut toujours supposer $\psi_n \geq 0$ sur $\partial\Omega$. Il résulte de (20) que

$$\|v(\psi_n) - v(\psi)\|_\infty \leq \|\psi_n - \psi\|_\infty$$

et donc $v(\psi_n)$ converge uniformément vers $v(\psi)$; il est clair alors comme ci-dessus que $v(\psi)$ est solution de $E(0,\psi)$. ∎

Donnons enfin la

<u>PREUVE DU LEMME 6.</u> Supposons d'abord (H2.b) et montrons que $D(A)$ est dense dans $C_o(\Omega)$. Utilisant Hahn-Banach, considérons $g \in C_o(\Omega)'$ telle que $\langle g,u\rangle = 0$ pour tout $u \in D(A)$, et montrons que $g = 0$. Puisque g est une mesure de Radon bonrnée sur Ω, il existe $g_n \in \mathcal{D}(\Omega)$ tels que

$$\int g_n u \to \langle g,u\rangle \text{ pour tout } u \in C_o(\Omega)$$

On peut toujours supposer A coercif sur $H^1_o(\Omega)$ en remplaçant éventuellement A par $A+\lambda$ avec une constante λ suffisamment grande; l'opérateur transposé

$$A^* = -\Sigma \frac{\partial}{\partial x_i}\left(\Sigma a_{ij} \frac{\partial}{\partial x_j} + a_i\right) + a_o$$

est alors aussi coercif sur $H^1_o(\Omega)$ et on peut considérer la solution $v_n \in H^1_o(\Omega)$ de $A^* v_n = g_n$ dans $\mathcal{D}'(\Omega)$.
On a

$$\int v_n Au = \int u g_n \to \langle g,u\rangle = 0 \text{ pour tout } u \in D(A)$$

et donc puisque $R(A) = L^\infty(\Omega)$ d'après le lemme 2,

$$v_n \to 0 \text{ dans } L^1(\Omega)$$

Maintenant A^{-1} est borné de $W^{-1,p}(\Omega)$ dans $L^\infty(\Omega)$ pour tout $p>N$ et donc par dualité, puisque $\{g_n\}$ est bornée dans $L^1(\Omega)$, on a $\{v_n\}$ bornée dans $W^{1,q}(\Omega)$ pour tout $q < \frac{N}{N-1}$ et donc

$$v_n \rightharpoonup 0 \text{ dans } W^{1,q}(\Omega) \quad \text{pour tout } q < \frac{N}{N-1}$$

Compte-tenu de l'hypothèse (H2.a), il en résulte que

$$g_n = A^* v_n \to 0 \quad \text{dans } \mathcal{D}'(\Omega)$$

et donc $g = 0$.

Pour prouver le lemme, donnons-nous $\tilde{\Omega}$ un ouvert borné régulier contenant $\bar{\Omega}$ et définissons

$$\tilde{a}_{ij}(x) = \begin{cases} a_{ij}(x) \\ \delta_{ij} \end{cases} \qquad \tilde{a}_i(x) = \begin{cases} a_i(x) & \text{si } x \in \Omega \\ 0 & \text{si } x \in \tilde{\Omega} \setminus \Omega \end{cases}$$

L'opérateur $\tilde{A}$ sur $\tilde{\Omega}$ correspondant vérifie (H1) et (H2).
Etant donné $\psi \in C(\bar{\Omega})$, on peut la prolonger en une fonction $\tilde{\psi} \in C_o(\tilde{\Omega})$.
D'après le paragraphe précédent, il existe $\tilde{u}_n \in D(\tilde{A})$ tels que $\tilde{u}_n \to \tilde{\psi}$ dans $C_o(\tilde{\Omega})$. Les restrictions u_n de $\tilde{u}_n$ à Ω sont bien dans $H^1(\Omega) \cap C(\bar{\Omega})$ avec $Au_n \in L^\infty(\Omega)$ et $u_n \to \psi$ dans $C(\bar{\Omega})$. ∎

Références

[1] L. Barthélemy, Thèse de 3ème cycle. Besançon (1980).

[2] L. Barthélemy et F. Catté, Annales Faculté des Sciences de Toulouse, Vol. IV (1982), pp. 165-190

[3] A. Bensoussan et J.L. Lions, Contrôle impulsionnel et Inéquations quasi-variationnelles. Dunod, Paris (1982).

[4] D. Gilbarg et N.S. Trudinge , Elliptic Partial Differential Equations of Second Order, Springer-Verlag (1977).

[5] B. Hanouzet et J.L. Joly, CRAS Paris série A, 286 (1978).

Louise Barthélemy & Philippe Bénilan
Equipe de Mathématique de Besançon
U.A. - C.N.R.S. 741
Université de Franche-Comté
25030 Besançon Cedex
France

J F BONNANS & E CASAS

Contrôle de systèmes elliptiques semilinéaires comportant des contraintes sur l'état

RESUME

Nous étudions le problème du contrôle optimal de systèmes gouvernés par des équations semilinéaires de type elliptique, lorsque l'état est soumis à des contraintes intégrales ou ponctuelles. Les conditions d'optimalité sont obtenues sous une forme non qualifiée. Des résultats de qualification sont donnés. En particulier, un lemme sur la stricte positivité des solutions lorsque le second membre est positif permet d'appliquer un critère général de qualification dans certains cas. L'emploi d'espaces adaptés permet d'étudier le cas des contraintes ponctuelles sur le gradient.

ABSTRACT

We study the control problem of systems governed by semi-linear elliptic equations, when the state is subject to integral or punctual constraints. The optimality conditions are obtained in a non-qualified form. Some qualification results are given. Specifically, a lemma on the strict positivity of solutions when the right-hand side of the equation is positive allows us to use a general criterion for qualification in some cases. The use of adapted spaces allows us to study the case of punctual constraints on the gradient.

I. Introduction

Des problèmes de contrôle de systèmes linéaires de type elliptique soumis à des contraintes sur l'état sont étudiés dans J.L. Lions [10], en l'absence de contraintes sur le contrôle. J. Mossino [11] étudie par une méthode de dualité le cas de contraintes ponctuelles sur le gradient de l'état, sans obtenir de conditions d'optimalité dans le cas où le contrôle est soumis à des contraintes. Le cas des contraintes ponctuelles sur l'état, en présence de contraintes sur le contrôle, est étudié par E. Casas [6], qui obtient l'existence du multiplicateur dans un espace de mesures et de l'état adjoint dans un espace du type $W^{1,s}(\Omega)$. Des résultats détaillés de régularité du contrôle optimal et de l'état associé sont analysés dans E. Casas [7]. Il est établi dans E. Casas [8] qu'une discrétisation par éléments finis de ce problème permet de calculer une approximation du contrôle optimal, et aussi du multiplicateur associé à la contrainte sur l'état.

Dans le cas de systèmes non linéaires, des résultats généraux sont obtenus dans J.F. Bonnans, E. Casas [3], et appliqués à des systèmes de type elliptique, parabolique ou hyperbolique. Nous reprenons ici des résultats généraux de [3] et les complétons dans le cas du contrôle distribué d'un système elliptique, en particulier par de nouveaux résultats de qualification. Par ailleurs, notons que le cas du système parabolique non linéaire est détaillé dans J.F. Bonnans, E. Casas [4], [5], où sont donnés des résultats de régularité du contrôle optimal.

Le papier est organisé comme suit. La partie II donne un résultat abstrait d'existence de multiplicateurs, sous une forme non qualifiée. Des critères de qualification sont donnés. La partie III donne des résultats d'existence d'un contrôle optimal pour un problème de contrôle d'un système elliptique. On envisage ensuite différents cas particuliers : le cas des contraintes intégrales est étudié en IV, celui des contraintes ponctuelles en V. Dans ces deux cas, la qualification du problème est obtenue en faisant certaines hypothèses sur les contraintes auxquelles est soumis le contrôle. Le cas des contraintes ponctuelles sur le gradient de l'état est étudié en VI par une méthode d'espaces adaptés.

II. Un résultat abstrait d'existence de multiplicateur

Donnons-nous

- X, W des espaces de Banach,
- K_1 un convexe fermé non vide de X,
- K_2 un convexe fermé d'intérieur non vide de W,
- f une application de X dans $\underline{R}$,
- g une application de X dans W,
- h une application de X dans $\underline{R}^n$, $n \in \underline{N}$ donné .

Nous rappelons qu'une application $g : X \to W$ est strictement dérivable en x si

$$\lim_{\substack{y \to x \\ t \to 0}} \frac{g(y+tz) - g(y)}{t} = Dg(x)z,$$

où $Dg(x) \in L(X,W)$, la convergence étant uniforme en y sur tout compact.

<u>REMARQUE 2.1</u>. Si g est continûment Gâteaux-différentiable en x, g est strictement dérivable en x (F.H. Clarke [9], p. 32). ∎

Considérons le problème

$$\text{Min } f(x), \; x \in K_1, \; g(x) \in K_2, \; h(x) \equiv 0. \tag{2.1}$$

Nous notons ∂ le sous-différentiel au sens de l'analyse convexe et ∂_c le gradient généralisé de F.H. Clarke [9].

<u>THEOREME 2.1</u>. Soit $\bar{x}$ une solution de (2.1). Nous supposons f et h lipschitziens et g strictement dérivable au voisinage de $\bar{x}$. Alors il existe $\alpha \geqslant 0$, $\lambda \in W'$ et $\mu \in \underline{R}^n$ tels que

$$\alpha + \|\lambda\|_{W'} + \sum_{i=1}^{n} |\mu_i| > 0, \tag{2.2}$$

$$\langle \lambda, w - g(\bar{x}) \rangle_{W'W} \leqslant 0, \; \forall w \in K_2, \tag{2.3}$$

$$\alpha \partial_c f(\bar{x}) + [Dg(\bar{x})]^* \lambda + \sum_{i=1}^{n} \mu_i \, \partial_c h_i(\bar{x}) + \partial I_{K_1}(\bar{x}) \ni 0 \text{ dans } X'. \; ∎ \tag{2.4}$$

<u>REMARQUE 2.2</u>. Naturellement, en l'absence de contraintes d'égalité, les résultats précédents restent valables en retirant les termes correspondant à ces contraintes.

Avant de démontrer le théorème, donnons un lemme dont on trouvera la démonstration dans J.F. Bonnans, E. Casas [3].

<u>LEMME 2.1</u>. Soit K un convexe fermé de W tel que $0 \in \overset{\circ}{K}$. Soit l'application $\rho : W \to \underline{R}$ définie par

$$\rho(w) = \inf\{\mu \geqslant 0;\ w \in \mu K\}.$$

Alors

$$\rho(tw) = t\,\rho(w),\ \forall t \geqslant 0,\ w \in W, \tag{2.5}$$

$$\rho(w) \leqslant 1 \Leftrightarrow w \in K, \tag{2.6}$$

$$\rho(w) < 1 \Leftrightarrow w \in \overset{\circ}{K},\ \rho(w) = 1 \Leftrightarrow w \in \partial K. \tag{2.7}$$

De plus ρ est sous-linéaire, convexe, lipschitzienne et

$$w \in \partial K \Rightarrow \partial\rho(w) \not\ni 0 \text{ et } \partial\rho(W) \subset \partial I_K(w). \tag{2.8}$$

Démonstration du théorème 2.1. Soit w_0 dans $\overset{\circ}{K}_2$. On applique le lemme 2.1 avec $K = K_2 - w_0$. En raison de (2.6), il vient

$$\rho(w - w_0) \leqslant 1 \Leftrightarrow w \in K_2.$$

Un problème équivalent à (2.1) est donc

$$\text{Min } f(x),\ x \in K_1,\ \rho(g(x) - w_0) - 1 \leqslant 0,\ h(x) \equiv 0. \tag{2.9}$$

Puisque f, ρ et h sont lipschitziennes au voisinage de $\bar{x}$, on déduit d'un résultat de F.H. Clarke (cf. [9]) que si $\bar{x}$ est solution de (2.1), donc de (2.9), il existe $\alpha \geqslant 0$, $\beta \geqslant 0$, $\mu \in \underline{R}^n$ tels que

$$\alpha + \beta + \sum_{i=1}^{n} |\mu_i| > 0,\ \beta = 0 \text{ si } \rho(g(\bar{x}) - w_0) < 1,$$

et

$$\alpha\partial_c f(\bar{x}) + \beta\partial_c(\rho(g(\bar{x}) - w_0)) + \sum_{i=1}^{n} \mu_i\, \partial_c\, h_i(\bar{x}) + \partial I_{K_1}(\bar{x}) \ni 0 \text{ dans } X'.$$

Or g(x) est strictement différentiable au voisinage de $\bar{x}$, et ρ est convexe et continue, donc régulière au sens de [9]; d'où (cf. [9], p. 45)

$$\partial_c(\rho(g(\bar{x}) - w_0)) = [D\, g(\bar{x})]^* \partial_c \rho(g(\bar{x}) - w_0).$$

De plus, ρ étant convexe et lipschitzienne, $\partial_c \rho$ coïncide avec le sous différentiel de ρ ([9], p. 36); donc il existe $\lambda \in \partial\rho(g(\bar{x}) - w_0)$ vérifiant (2.4), avec $\lambda = 0$ si $g(\bar{x}) \in \overset{\circ}{K}_2$. Si $g(\bar{x}) \in \partial K$, $\lambda \in \partial\rho(g(\bar{x}) - w_0)$ est d'après (2.8) un élément non nul de ∂I_K en $g(\bar{x}) - w_0$, donc de ∂I_{K_2} en $g(\bar{x})$; d'où (2.3) et, avec (2.10), (2.2). La conclusion du théorème s'ensuit. ■

Il est utile d'obtenir des conditions sous lesquelles le problème est normal, au sens où la conclusion du théorème 2.1 est obtenue avec $\alpha = 1$. Nous allons donner deux indications dans ce sens; l'une est du type de la condition de Slater; l'autre est obtenue en considérant un problème obtenu par une perturbation de K_2, et utilise les résultats de F.H. Clarke [9]. Les résultats nécessitent en particulier des hypothèses supplémentaires sur h.

THEOREME 2.2.

(i) Si h est strictement dérivable, si $\{\nabla h_i(\bar{x})\}$ est linéairement indépendant et s'il existe $x_0 \in \overset{\circ}{K}_1$ (K_1 si aucune contrainte d'égalité n'est présente) tel que

$$\begin{aligned} &(i)\ g(\bar{x}) + Dg(\bar{x})(x_0 - \bar{x}) \in \overset{\circ}{K}_2, \\ &(ii)\ \langle \nabla h_i(\bar{x}), x_0 - \bar{x} \rangle = 0,\ i = 1 \text{ à } n, \end{aligned} \tag{2.11}$$

la conclusion du théorème 3.1 est obtenue avec $\alpha = 1$.

(ii) Supposons qu'aucune contrainte d'égalité n'est présente. Soit w_0 dans $\overset{\circ}{K}_2$. Considérons pour $\gamma \in \underline{R}^+$ le convexe perturbé

$$K_{2\gamma} = w_0 + \gamma(K_2 - w_0),$$

et soit la famille de problèmes

$$\text{Min } f(x);\ x \in K_1,\ g(x) \in K_{2\gamma}, \tag{2.12}$$

qui se réduit à (2.1) lorsque $\gamma = 1$. Soit a,b tels que $0 \leq a < b \leq +\infty$ et que le problème (2.12) possède au moins une solution si $\gamma \in [a,b[$. Alors, pour presque tout $\gamma \in [a,b[$, le problème (2.12) est normal. ■

Démonstration

(i) Si α est nul, nous déduisons de (2.4) et (2.11ii) que, en particulier :

$$<\lambda, Dg(\bar{x}) (x_0 - \bar{x})>_{W'W} \geq 0. \tag{2.13}$$

D'autre part, il existe $\varepsilon > 0$ tel que

$$\|w\|_W < \varepsilon \Rightarrow g(\bar{x}) + Dg(\bar{x}) (x_0 - \bar{x}) + w \in \mathring{K}_2 ,$$

donc avec (2.3)

$$<\lambda, Dg(\bar{x}) (x_0 - \bar{x}) + w>_{W'W} \leq 0, \ \forall w \in W, \ \|w\|_W \leq \varepsilon,$$

Si $\|\lambda\|_{W'} \neq 0$ (ce sera le cas d'après (2.2) si aucune contrainte d'égalité n'est présente) il s'ensuit que

$$<\lambda, Dg(\bar{x}) (x_0 - \bar{x})>_{W'W} < 0,$$

en contradiction avec (2.13). Si $\|\lambda\|_{W'} = 0$, on déduit de (2.2) que μ n'est pas nul et de (2.4) que pour un certain $q \in \partial I_{K_1} (\bar{x})$:

$$\sum_{i=1}^{n} \mu_i \nabla h_i(\bar{x}) + q = 0.$$

Donc q est non nul et $<q, x_0 - \bar{x}> < 0$ puisque $x_0 \in \mathring{K}_1$, ce qui est impossible d'après la relation ci-dessus.

(ii) Le problème (2.12) équivaut à

$$\begin{aligned} &\text{Min } f(x), \\ &x \in K_1, \ \rho(g(\bar{x}) - w_0) \leq \gamma. \end{aligned} \tag{2.14}$$

Soit x_γ une solution de (2.14) et soit $\phi : [a,b[\to \underline{R}$ définie par

$$\phi(\gamma) = f(x_\gamma).$$

La fonction ϕ est décroissante en finie en tout point. On déduit alors de F.H. Clarke [14], la normalité de (2.14) pour presque tout γ. En reprenant la démonstration du théorème 4.1, on vérifie que la normalité de (2.12) équivaut à celle de (2.14), d'où le théorème. ■

III. Controle d'un système non linéaire de type elliptique comportant des contraintes ponctuelles sur l'état

Soit Ω un ouvert connexe de $\underline{R}^n$, $n \leq 3$, de frontière Γ régulière. Soit $f : R \to \underline{R}$ de classe C^1, monotone croissante. Considérons le système

$$\begin{aligned} &- \Delta y + f(y) = u \text{ dans } \Omega, \\ &y = 0 \text{ sur } \Gamma . \end{aligned} \tag{3.1}$$

Le résultat suivant est bien connu :

PROPOSITION 3.1. Pour tout $u \in L^2(\Omega)$, l'équation (3.1) admet une solution unique y_u dans $H^2(\Omega) \cap H_0^1(\Omega)$. ■

Notons l'inclusion $H^2(\Omega) \overset{n\leq 3}{\subset} C(\bar{\Omega})$. L'état y_u est donc une fonction continue de x. Soit $N \geq 0$ et y_d dans $L^2(\Omega)$. Posons

$$J(u) = \frac{N}{2} \int_\Omega (u(x))^2 \, dx + \frac{1}{2} \int_\Omega (y_u(x) - y_d(x))^2 \, dx. \tag{3.2}$$

Soit K (resp. K_Y) un ensemble convexe fermé de $L^2(\Omega)$ (resp. $H^2(\Omega) \cap H_0^1(\Omega)$). Considérons le problème

$$\min J(u); \ u \in K, \ y_u \in K_Y. \tag{3.3}$$

On obtient par le procédé habituel de passage à la limite dans une suite minimisante le

THEOREME 3.1. Si'il existe $u_0 \in K$ tel que $y_{u_0} \in K_Y$, et si

$N > 0$ ou K est un borné de $L^2(\Omega)$,

alors le problème (3.3) admet (au moins) un contrôle optimal. ■

En vue d'exprimer les conditions nécessaires d'optimalité, donnons deux resultats concernant la dépendance de l'état par rapport au contrôle.

PROPOSITION 3.2. L'application $u \to y_u$ est de classe C^1 de $L^2(\Omega)$ vers $H^2(\Omega) \cap H_0^1(\Omega)$ et sa dérivée z dans la direction v est solution de

$$\begin{aligned} & -\Delta z + f'(y_u) z = v \text{ dans } \Omega, \\ & z = 0 \text{ sur } \Gamma. \end{aligned} \tag{3.5}$$

Cette proposition se démontre en utilisant le théorème des fonctions implicites.

PROPOSITION 3.3. Soit u, v dans $L^2(\Omega)$ tels que $u \geqslant v$ et $u \neq v$. Alors $y_u > y_v(x)$, $\forall x \in \Omega$. ■

Démonstration

Posons

$$a(x) = \begin{cases} (f(y_u(x)) - f(y_v(x)))/(y_u(x) - y_v(x)) & \text{si } y_u(x) \neq y_v(x), \\ f'(y_u(x)) & \text{sinon.} \end{cases}$$

La continuité de y_u, y_v et de f' impliquent celle de $a(x)$. De plus, $a(x)$ est positive et $w = y_u - y_v$ vérifie

$$\begin{aligned} & -\Delta w + a(x) w = u - v \text{ dans } \Omega, \\ & w = 0 \text{ sur } \Gamma. \end{aligned}$$

La positivité stricte de w (qui donne la conclusion) est une conséquence du lemme suivant. ■

LEMME 3.1. Soit $a(x)$ (resp. $u_0(x)$) un élément positif de $L^\infty(\Omega)$. On suppose u_0 non identiquement nul. Alors l'équation

$$\begin{aligned} &-\Delta w + a(x)w = u_0 \text{ dans } \Omega, \\ &w = 0 \text{ sur } \Gamma, \end{aligned} \tag{3.6}$$

admet une solution w, continue, strictement positive sur Ω.

Démonstration

On sait que (3.6) admet une solution w continue, positive, non identiquement nulle. Soit $x_0 \in \Omega$ tel que $w(x_0) > 0$ et supposons l'existence de $x_1 \in \Omega$ tel que $w(x_1) = 0$. Soit $\Omega_1 \subset \Omega$ un ouvert connexe de frontière Γ_1 régulière telle que $x_0 \in \Gamma_1$ et $x_1 \in \Omega_1$. Soit w_1 la solution dans $H^2(\Omega_1)$ de

$$\begin{aligned} &-\Delta w_1 + a(x)\, w_1 = 0 \text{ dans } \Omega_1, \\ &w_1 = w \text{ sur } \Gamma_1. \end{aligned} \tag{3.7}$$

D'après un corollaire de l'inégalité de Harnack (G. Stampacchia [12]) w_1 est identiquement nul ou strictement positif sur Ω_1. Mais il ne peut être nul d'après la condition aux limites de (3.7). La contradiction est obtenue en utilisant le principe du maximum qui implique que $w_1 \leqslant w$ sur Ω_1 et en particulier $w_1(x_1) \leqslant w(x_1) = 0$. ■

Etudions maintenant les conditions d'optimalité en considérant quelques cas particuliers de contraintes sur l'état.

IV. Le cas des contraintes intégrales sur l'état

Nous supposons ici que, pour $\delta > 0$ donné :

$$K_Y = \{y \in H^2(\Omega) \cap H_0^1(\Omega),\ \int_\Omega |y(x)|\, dx \leqslant \delta\}.$$

Pour éviter d'utiliser l'espace dual de $H^2(\Omega) \cap H_0^1(\Omega)$, posons

$$B_\delta = \{z \in L^1(\Omega)\ ;\ \|z\|_{L^1(\Omega)} \leqslant \delta\} \quad .$$

Le problème de contrôle peut alors s'écrire

$$\text{Min } J(u) \ ; \ u \in K, \ y_u \in B_\delta. \tag{4.1}$$

Les hypothèses du théorème 2.2 sont vérifiées avec $X = L^2(\Omega)$, $W = L^2(\Omega)$, $K_1 = K$, $K_2 = B_\delta$. On en déduit le

THEOREME 4.1. Pour toute solution $\bar{u}$ du problème (4.1), il existe $\bar{\mu}$ dans $L^\infty(\Omega)$, $\bar{p}$ et $\bar{q}$ dans $H^2(\Omega) \cap H_0^1(\Omega)$ et $\alpha \geq 0$ tels que (nous notons $\bar{y} = y_{\bar{u}}$) :

$$\alpha + \|\bar{p}\| > 0, \ \int_\Omega \bar{\mu}(z - \bar{y})dx \leq 0, \ \forall z \in B_\delta,$$

$$-\Delta\bar{p} + f'(\bar{y})p = \bar{\mu} \text{ dans } \Omega, \qquad -\Delta\bar{q} + f'(\bar{y})\bar{q} = \bar{y} - y_d \text{ dans } \Omega,$$

$$\bar{p} = 0 \text{ sur } \Gamma, \qquad \bar{q} = 0 \text{ sur } \Gamma,$$

et

$$\int_\Omega [\bar{p} + \alpha(N\bar{u} + \bar{q})](v - \bar{u}) \, dx \geq 0, \ \forall v \in K. \ \blacksquare$$

Démonstration

On sait que l'équation en $\bar{q}$ est bien posée et qu'elle fournit l'expression du gradient du critère. On applique alors le théorème 2.1 avec $g(u) = y_u$, en notant que la non nullité de $\bar{\mu}$ équivaut à celle de $\bar{p} = [\partial g/\partial u(\bar{u})]^* \bar{\mu}$. Enfin on interprète $\bar{p}$ comme la solution d'un problème aux limites. ■

REMARQUE 4.1.
(i) Soit s l'application de $\underline{R} \to 2^{\underline{R}}$ définie par

$$s(t) = \begin{cases} -1 & \text{si } t < 0, \\ [-1,+1] & \text{si } t = 0, \\ 1 & \text{si } t > 0. \end{cases}$$

On sait qu'il existe $\lambda > 0$ tel que $\bar{\mu}(x) \in \lambda s(y(x))$, p.p. $x \in \Omega$.

(ii) D'après les résultats de [2], $\bar{p}$ est dans $W^{2,s}(\Omega)$ pour tout $s \geq 1$. ■

Intéressons nous maintenant à la question de la qualification du problème

($\alpha = 1$).

THEOREME 4.2.

(i) Supposons l'hypothèse (3.4) vérifiée ainsi que l'existence d'un contrôle admissible pour $\delta = \delta_0$. Alors le problème (4.1) admet une solution pour tout $\delta \geqslant \delta_0$ et il est qualifié ($\alpha = 1$) pour presque tout $\delta \geqslant \delta_0$.

(ii) Supposons f (donc l'équation d'état) linéaire. Alors si $0 \in K$, le problème est qualifié.

(iii) Si $K = L^2(\Omega)$ (pas de contraintes sur le contrôle) le problème est qualifié. ∎

Démonstration

(i) On applique le point (ii) du théorème 2.2.

(ii) On applique le point (i) du théorème 2.2.

(iii) Si $\alpha = 0$, on déduit de (4.2) que $\bar{p} = 0$, en contradiction avec la conclusion du théorème. ∎

Naturellement, la qualification du problème est difficile à vérifier si on a à la fois une équation d'état non linéaire et des contraintes sur le contrôle. Voici un résultat partiel dans ce sens.

THEOREME 4.3. On suppose que $f(0) = 0$ et

$$K = \{u \in L^2(\Omega);\ 0 \leqslant u(x) \leqslant \beta, \text{ p.p. } x \in \Omega\} ,$$

avec $0 < \beta \leqslant +\infty$. Alors le problème (4.1) est qualifié. ∎

Démonstration

Si $\bar{u} = 0$, le résultat est trivial; sinon, d'après la proposition 3.3, on a $\bar{y}(x) > 0$, $\forall x \in \Omega$. On en déduit que $\bar{\mu}$ est une constante positive (cf. remarque 2.1.i), strictement positive si $\alpha = 0$. Dans ce cas, d'après le lemme 3.1, $\bar{p}$ est strictement positif sur Ω. De (4.2) on déduit alors que $\bar{u} = 0$ d'où une contradiction. ∎

REMARQUE 4.2. Si le problème est qualifié ($\alpha = 1$) et si $N > 0$ on déduit de (4.2) que $\bar{u} = P_K(-\frac{1}{N}(\bar{p} + \bar{q}))$. Ceci, joint à des hypothèses sur K, permet d'obtenir des résultats de régularité du contrôle optimal. Ainsi, dans l'exemple précédent, $\bar{u}$ est dans $H^1(\Omega) \cap C(\bar{\Omega})$ et y est dans $H^3(\Omega)$. Pour d'autres exemples, voir E. Casas [7]. ∎

V. Le cas des contraintes ponctuelles sur l'état

Envisageons maintenant le cas où, pour $\delta > 0$ donné

$$K_Y = \{Y \in H^2(\Omega) \cap H_0^1(\Omega) ; \ |y(x)| \leq \delta, \ \forall x \in \Omega\}.$$

Soit $C_0(\bar{\Omega})$ l'espace des fonctions continues sur $\bar{\Omega}$, nulles sur Γ, muni de la norme du maximum. Posons

$$B_\delta = \{z \in C_0(\bar{\Omega}), \ |z(x)| \leq \delta, \ \forall x \in \Omega\},$$

et écrivons le problème de contrôle comme

$$\text{Min } J(u) ; \ u \in K, \ y_u \in B_\delta. \tag{5.1}$$

Notons $M(\Omega)$ l'espace des mesures de Borel réelles et régulières sur Ω. C'est le dual de $C_0(\bar{\Omega})$. L'application du théorème 2.1 avec $X = L^2(\Omega)$, $W = C_0(\bar{\Omega})$, $K_1 = K$, $K_2 = B_\delta$ donne le

THEOREME 5.1. A toute solution $\bar{u}$ du problème (5.1) sont associés $\bar{\mu}$ dans $M(\Omega)$, $\bar{p}$ dans $W_0^{1,s}(\Omega)$ pour tout $s \in [1, n/(n-1)[$, $\bar{q}$ dans $H^2(\Omega) \cap H_0^1(\Omega)$, et $\alpha \geq 0$ tels que

$$\alpha + \|\bar{p}\|_{W_0^{1,1}(\Omega)} > 1 \ ; \tag{5.2}$$

$$\int_\Omega (z - \bar{y}) d\bar{\mu} \leq 0, \ \forall z \in B_\delta, \tag{5.3}$$

$$\begin{aligned} -\Delta\bar{p} + f'(\bar{y})\bar{p} &= \bar{\mu} \text{ sur } \Omega, \\ \bar{p} &= 0 \text{ sur } \Gamma, \end{aligned} \tag{5.4}$$

$$-\Delta\bar{q} + f'(\bar{y})\bar{q} = \bar{y} - y_d \text{ sur } \Omega,$$
$$\bar{q} = 0 \text{ sur } \Gamma, \tag{5.5}$$

$$\int_\Omega [\bar{p} + \alpha(N\bar{u} + \bar{q})](v-\bar{u})dx \geq 0, \ \forall v \in K. \ \blacksquare \tag{5.6}$$

Démonstration

Elle est identique à celle du théorème 4.1. On sait que $\bar{p} \in L^2(\Omega)$ et $-\Delta\bar{p} \in M(\Omega)$; la régularité de $\bar{p}$ se déduit alors de E. Casas [7] ou G. Stampacchia [12]. ■

REMARQUE 5.1. La décomposition de Jordan de μ en $\mu^+ - \mu^-$ vérifie : μ^+ (resp. μ^-) a pour support $\{x \in \Omega,\ y(x) = 1 \text{ (resp. } -1)\}$ (voir à ce sujet E. Casas [7]).

Etudions maintenant la question des conditions de qualification. L'analogue du théorème 4.2 s'obtient sans difficulté. Nous reprenons le cas où K est donné par (4.2).

THEOREME 5.2. Nous supposons que $f(0) = 0$ et que K est donné par (4.2), avec $0 < \beta \leq +\infty$. Alors le problème (5.1) est qualifié. ■

Démonstration

Si $\bar{u} = 0$, le résultat est vrai. Sinon, soit z la solution du problème

$$-\Delta z + f'(\bar{y})z = \bar{u} \text{ sur } \Omega,$$
$$z = 0 \text{ sur } \Gamma.$$

D'après le lemme 3.1, $z(x) > 0$ pour tout x de Ω. Comme $\bar{y}$ est positif, on en déduit que, pour $\varepsilon > 0$ assez petit, $|\bar{y}(x) - \varepsilon z(x)| < 1$ pour tout x de Ω, d'où :

$$\exists \gamma < 1; \ |\bar{y}(x) - \varepsilon z(x)| \leq \gamma, \ \forall x \in \Omega.$$

On peut alors appliquer le point (i) du théorème 2.2, avec $v = (1-\varepsilon)\bar{u}$, qui donne le résultat. ■

REMARQUE 5.2. Dans l'exemple précédent, de la qualification du problème et de (5.6) on déduit que si $N > 0$, $\bar{u}$ est dans $W_0^{1,s}(\Omega)$ pour $s < n/(n-1)$, donc $\bar{y}$ est dans $W^{3,s}(\Omega)$ d'après [2]. ■

VI. Le cas des contraintes ponctuelles sur le gradient

Nous supposons ici que, $\| \ \|$ étant une norme donnée de $\mathbb{R}^n$:

$$K_Y = \{y \in H^2(\Omega) \cap H_0^1(\Omega)\}; \quad \|\nabla y(x)\| \leqslant \delta, \text{ p.p. } x \in \Omega\}. \tag{6.1}$$

Considérons donc

$$B_\delta = \{z \in L^\infty(\Omega)^n, \ \|z(x)\| \leqslant \delta, \text{ p.p. } x \in \Omega\} ,$$

et écrivons le problème de contrôle sous la forme

$$\min J(u); \ u \in K, \ \nabla y_u \in B_\delta .$$

La difficulté vient ici de ce que l'application $u \to \nabla y_u$ n'est pas continue de $L^2(\Omega)$ vers $L^\infty(\Omega)^n$. Il est donc nécessaire de choisir un cadre fonctionnel spécialement adapté au problème.

Pour tout u de $L^2(\Omega)$, appelons z_u la solution de

$$-\Delta z = u \text{ dans } \Omega,$$
$$z = 0 \text{ sur } \Gamma.$$

Pour que l'application $u \to \nabla y_u$ soit de classe C^1, à image dans $L^\infty(\Omega)^n$, on introduit les espaces adaptés

$$Y = \{y \in H^2(\Omega) \cap H_0^1(\Omega); \ \nabla y \in L^\infty(\Omega)^n\},$$

$$U = \{u \in L^2(\Omega) \ ; \ z_u \in Y\}.$$

Munis de la norme du graphe, Y et U sont des espaces de Banach.

REMARQUE 6.1. Pour $p > n$, $L^p(\Omega) \subset U$; donc U est dense dans $L^2(\Omega)$. ∎

Les espaces U et Y sont bien adaptés à notre problème, en effet :

LEMME 6.1. On a

$$\{u \in U\} \Leftrightarrow \{\nabla y_u \in L^\infty(\Omega)^n\},$$

et l'application $u \to y_u$ est de classe C^1 de U vers $L^\infty(\Omega)^n$. ∎

Démonstration

Soit $w = z_u - y_u$. Il vient

$$-\Delta w = f(y_u) \text{ dans } \Omega,$$
$$w = 0 \text{ sur } \Gamma.$$

Puisque $f(y_u)$ est continue, w est donc dans $W^{2,s}(\Omega)$ pour tout $s \in [1,+\infty[$, d'où $\nabla w \in W^{1,s}(\Omega)^n \overset{s>n}{\subset} C(\bar{\Omega})^n$. La régularité de l'application $u \to y_u$ s'obtient en appliquant le théorème des fonctions implicites à

$$F : Y \times U \to U,$$
$$(y,u) \to -\Delta y + f(y) - u.$$

On vérifie que F est de classe C^1. Il faut ensuite établir l'existence d'une solution unique dans Y, de l'équation

$$-\Delta z + f'(\bar{y})z = v \text{ dans } \Omega,$$
$$z = 0 \text{ sur } \Gamma.$$

L'appartenance de Δz à U implique celle de z à Y. ∎

Appliquons maintenant le théorème 2.1 au problème

$$\min J(u) \; ; \; u \in K \cap U, \; \nabla y_u \in B_\delta. \tag{6.3}$$

Nous en déduisons le

THEOREME 6.1. Si l'hypothèse (6.1) est vérifiée, à toute solution $\bar{u}$ de (6.3) sont associées $\mu \in [L^\infty(\Omega)^n]'$, $\bar{p} \in U'$, $\bar{q} \in H^2(\Omega) \cap H_0^1(\Omega)$, et $\alpha \geq 0$ tels que

$$\alpha + \|\bar{p}\|_{U'} > 0; \tag{6.4}$$

$$<\bar{\mu}, z - \nabla\bar{y}> \leq 0, \quad \forall z \in B_\delta, \tag{6.5}$$

$$<\bar{p}, \quad -\Delta z \ + f'(\bar{y})z>_{U'U} = <\bar{\mu}, \nabla z>, \quad \forall z \in Y, \tag{6.6}$$

$$\begin{aligned} & -\Delta\bar{q} + f'(\bar{y})\bar{q} = \bar{y} - y_d \text{ dans } \Omega, \\ & \bar{q} = 0 \text{ sur } \Gamma, \end{aligned} \tag{6.7}$$

et

$$<\bar{p}, \ v - \bar{u}> + \alpha \int_\Omega (N\bar{u} + \bar{q})(v - \bar{u})dx \geq 0, \quad \forall v \in K \cap U. \quad \blacksquare$$

Démonstration

Le seul point délicat est de montrer que $\bar{p} \neq 0$ si $\bar{\mu} \neq 0$. Il suffit de prendre $z = \bar{y}$ dans (6.6) : en effet, de (6.5) on déduit que $<\bar{\mu}, \nabla\bar{y}> > 0$.

REMARQUE 6.2.

(i) L'équation (6.6) définit $\bar{p}$ de façon unique (car l'équation d'état linéarisée est une isomorphisme de U dans Y).

(ii) Si $K \subset L^s(\Omega)$, $s > n$, il vaut mieux choisir $L^s(\Omega)$ comme espace des contrôles : ceci permet d'obtenir $\bar{p}$ dans $L^{s'}(\Omega)$ avec $1/s + 1/s' = 1$, et $\bar{\mu}$ dans $M(\bar{\Omega})^n$. Ceci vaut en particulier si on impose que $|u(x)| \leq M$, p.p. $x \in \Omega$.

(iii) On vérifie comme précédemment que le problème est qualifié si $K = L^2(\Omega)$ ou si f est linéaire et K contient $\{0\}$. Si le problème est non linéaire et comporte des contraintes sur le contrôle et l'état, la question de la qualification est ouverte. Notons toutefois qu'on peut appliquer le point (ii) du théorème (2.2) pour obtenir des résultats de qualification pour presque tout δ. ■

References

[1] R.A. Adams, Sobolev spaces. Academic Press, New York (1975).

[2] S. Agmon, A. Douglis, L. Nirenberg, Estimates near the boundary for the solutions of elliptic partial differential equations satisfying general boundary conditions. Comm. on Pure and Appl. Math. 12, p. 632-727 (1959).

[3] J.F. Bonnans, E. Casas, Contrôle de systèmes non linéaires comportant des contraintes distribuées sur l'état, Rapport INRIA no 300 (1984).

[4] J.F. Bonnans, E. Casas, On the choice of the function spaces for some state-constrained control problems, Numer. Funct. Anal. Optim. 7, p. 333 - 348 (1984-1985).

[5] J.F. Bonnans, E. Casas, Quelques méthodes pour des problèmes de contrôle avec des contraintes sur l'état. Ann. Scient. Univ. Al. I. Cuza XXXII, p. 57-62 (1986).

[6] E. Casas, Quelques problèmes de contrôle avec contraintes sur l'état. C.R. Acad. Sc. Paris, 296, p. 509-512 (1983)

[7] E. Casas, Control of an elliptic problem with pointwise state constraints. SIAM J. Cont. Optim. 24, p. 1309-1318 (1986).

[8] E. Casas, Un problema de control de sistemas gobernados por ecuaciones en derivadas parciales. V.C.E.D.Y.A., Tenerife (1982).

[9] F.H. Clarke, Optimization and nonsmooth analysis. Wiley, New York (1983).

[10] J.L. Lions, Contrôle optimal de systèmes gouvernés par des équations aux dérivées partielles. Paris, Dunod (1968).

[11] J. Mossino, An application of duality to distributed optimal control problems with constraints on the control and the state. J. of Math. an Appl. 50, pp. 223-242 (1975).

[12] G. Stampacchia, Le problème de Dirichlet pour les équations elliptiques du second ordre à coefficients discontinus. Ann. Inst. Fourier 15, p. 189-258 (1965).

Joseph F. Bonnans
INRIA
Domaine de Voluceau
B.P. 105
Rocquencourt
78153 Le Chesnay Cedex
France

Eudardo Casas
Departamento de Ecuaciones Funcionales
Facultad de Ciencias
Universidad de Santander
39005 Santander
Espagne

F H CLARKE

Action principles and periodic orbits

> "Then he told me something which I found absolutely fascinating,and have, since then, always found fascinating. Every time the subject comes up, I work on it. In fact, when I began to prepare this lecture I found myself making more analyses on the thing. Instead of worrying about the lecture, I got involved in a new problem. The subject is this - the principle of least action."
>
> Richard Feynman (The Feynman Lectures)

Principles of least action, and periodic trajectories : these are our two topics in this article. Each topic has a vast literature, so that a comprehensive survey is out of the question. To give this exposition a clear purpose, we shall focus upon a specific technique, namely the use of actual optimization of variational functionals in the study of periodicity.

1. Three action integrals

We begin by presenting the three different actions that will be discussed in the article.

1.1. The Lagrangian action

In what surely ranks among the most far-reaching mathematical works of all time, Leonhard Euler in 1744 laid the foundations for the calculus of variations and introduced the technique of variational principles. (Goldstine [16] devotes a chapter to discussing Euler's monograph). He focused upon the basic problem in the calculus of variations, that of minimizing the functional $A_L(x)$ defined by

$$\int_0^T L(t,x(t),\dot{x}(t))dt \qquad (1)$$

where L (the *Lagrangian*) is a given function mapping $[0,T] \times \mathbb{R}^n \times \mathbb{R}^n$ to $\mathbb{R}$, and where the minimization takes place over some class of functions X mapping $[0,T]$ to $\mathbb{R}^n$. (The notation $\dot{x}$ refers to dx/dt.) In this section we shall simplify the presentation by assuming that all the functions involved have all necessary derivatives; we shall also ignore for the time being the *endpoint constraints* that may be imposed on the values x(0) and x(T).

Euler derived the necessary condition that must be satisfied by a function x which minimizes the functional A_L :

$$\frac{d}{dt} L_v(t,x(t),\dot{x}(t)) = L_x(t,x(t),\dot{x}(t)) \tag{2}$$

This is called the *Euler equation*. Of equal concern to us here is that Euler went on to enunciate the principle that the behavior of a physical system (i.e. its equations of motion) could be described as corresponding to the minimization of a certain functional A_L related to the system; the functional is then referred to as the *action*. The canonical example of this "principle of least action" is the case of a single particle acted upon by a conservative force field. Denoting the position of the particle by x (in $\mathbb{R}^3$) and the potential energy by V(x), the corresponding action A_L is defined via

$$A_L(x) = \int_0^T \{\frac{m|\dot{x}|^2}{2} - V(x)\}dt \tag{3}$$

It is an easy matter to verify that the Euler equation for the functional (3) reduces to $m\ddot{x} = -\nabla V(x)$, which is Newton's equation. We recognize the first term in the integrand as the kinetic energy.

The action principle remains valid when constraints on the particle lead to the use of generalized coordinates to describe its position. A specific example of this, and one that we shall discuss at length in Section 5, is the *nonlinear pendulum*. Letting x denote the angular displacement from the (lower) rest position, the potential energy is seen to be $mg\ell(1-\cos x)$, where ℓ is the length of the pendulum. The action is

$$m\ell^2 \int_0^T \{\dot{x}^2/2 + (g/\ell)\cos x - g/\ell\}dt \tag{4}$$

There may very well be more than one action corresponding to a given situation, just as there may also be different sets of variables that can be used to describe it, as we now see.

1.2. The Hamiltonian formulation

Given a Lagrangian L, let us now consider the algebraic equation

$$p = L_v(t,x,v),$$

and assume that it defines v as a function of (t,x,p). We then proceed to define a function $H : [0,T] \times \mathbb{R}^n \times \mathbb{R}^n \to \mathbb{R}$ (the *Hamiltonian*) via

$$H(t,x,p) = p.v(t,x,p) - L(t,x,v(t,x,p)). \tag{5}$$

(This is known as the *Legendre transform*.) It is an easy exercise to show that the Euler equation for L can be expressed in an equivalent way in terms of H as follows : define

$$p(t) = L_v(t,x(t),\dot{x}(t)).$$

Then x satisfies the Euler equation iff (x,p) satisfies the following system of Hamiltonian equations :

$$\begin{aligned} -\dot{p}(t) &= H_x(t,x(t),p(t)) \\ \dot{x}(t) &= H_p(t,x(t),p(t)) \end{aligned} \tag{6}$$

Hamiltonian systems play the central role in analytical mechanics. Many developments of the classical theory are couched in Hamiltonian terms : Noether's theorem, canonical transformations, perturbation analysis, the Hamilton-Jacobi equation, etc. When H is independent of t, it follows easily from (6) that H is constant along the trajectory (x(t),p(t)); this constant is usually called the *energy*. In the canonical example of a particle in a conservative force field, we find $p = m\dot{x}$, the momentum, and H = potential + kinetic energy = total energy.

It is possible to define an action principle (*Hamilton's principle*) in terms of H and the action functional $A_H(x,p)$ defined by

$$A_H(x,p) = \int_0^T \{\dot{p}.x + H(t,x,p)\}dt \qquad (7)$$

Note that this action deals with two free functions x and p taking values in $\mathbb{R}^n$, in contrast to $A_L(x)$. The Euler equation for A_H is seen to be

$$d/dt\ [0,x] = [p + H_x,\ H_p],$$

which is precisely the Hamiltonian system (6). Thus A_H is a "valid" action; its extremals (solutions of the Euler equation) are Hamiltonian trajectories.

We see therefore that the Hamiltonian formulation is just a restatement of the Lagrangian theory. Yet there can be excellent reasons for preferring one action to another, as we shall see. But first we introduce our final action functional.

1.3. The dual action

The dual action is defined in terms of a new function G which we now consider along with L and H. The provenance of G is best described in terms of what is known as the *Fenchel transform* in convex analysis; we proceed to describe this in general terms.

Let f be a function mapping $\mathbb{R}^m$ to $R \cup \{+\infty\}$. Its *conjugate* (or Fenchel transform) is the extended-valued function f^* defined by

$$f^*(\zeta) = \sup\ \{\zeta.z - f(z)\ :\ z \in \mathbb{R}^m\} \qquad (8)$$

This operation is an involution on the class of lower-semicontinuous convex functions f; i.e. we then have $f^{**} = f$. As an example of conjugacy, one pertinent later, we remark that $(z^p/p)^* = (\zeta^q/q)$, if p and q satisfy $1/p + 1/q = 1$. The main fact regarding conjugates that we shall need is the inversion of derivatives. When both f and f^* are differentiable, for example, this asserts

$$\zeta = \nabla f(z) \quad \text{iff } z = \nabla f^*(\zeta). \qquad (9)$$

In general, the inversion relation involves the *subdifferentials* (see [2] or [22]) of the (not necessarily differentiable) functions f and f^* :

$$\zeta \in \partial f(z) \quad \text{iff } z \in \partial f^*(\zeta). \tag{9}^*$$

The set $\partial f(z)$ reduces to $\{\nabla f(z)\}$ when f is differentiable at z. We defer further comment about generalized gradients to the next section.
When f is convex and exhibits sufficient growth, the supremum in (8) is attained at a unique point z, namely the one satisfying $\nabla f(z) = \zeta$; then $f^*(\zeta)$ is just $\zeta.z - f(z)$. This is precisely the operation defining the Legendre transform. We see therefore that under certain mild conditions on L (rarely absent in practice) we have the following alternate characterization of the Hamiltonian :

$$H(t,x,p) = \sup \{p.v - L(t,x,v) \ : \ v \in \mathbb{R}^n\} \tag{10}$$

(i.e. $H(t,x,.) = L(t,x,.)^*$). This is the definition we shall employ henceforth; note that it does not require that L be differentiable.
We now proceed to define the function G mentioned earlier :

$$G(t,y,q) = \sup \{(y,q).(x,p) - H(t,x,p) \ : \ (x,p) \in \mathbb{R}^n \times \mathbb{R}^n\} \tag{11}$$

That is, in terms of conjugates, $G(t,.,.) = H(t,.,.)^*$. Clearly under suitable hypotheses this transformation could be defined in terms similar to the Legendre transform. It is also possible to define G directly from L by substituting (10) into (11) :

$$G(t,y,q) = \sup_{x,p} \ \inf_v \{(y,q).(x,p) - p.v + L(t,x,v)\}.$$

Under mild hypotheses the supremum and infimum can be switched, which leads to the expression

$$G(t,y,q) = \sup_x \ \{y.x + L(t,x,q)\} \ .$$

Comparison of this with (10) shows that G may be thought of in terms similar to H, but with the transformation occurring in the x rather than the v variable. We shall refer to G as the *dual Hamiltonian*.

We now define the *dual action* A_G in terms of G as follows

$$A_G(x,p) = \int_0^T \{\dot{p}.x + G(t,-\dot{p},\dot{x})\}dt \tag{12}$$

For the canonical example (3), G(y,q) turns out to be $V^*(y) + m|q|^2/2$, and the dual action is given by

$$\int_0^T \{\dot{p}.x + V^*(-\dot{p}) + \frac{m|\dot{x}|^2}{2}\}dt \tag{13}$$

The dual action is of considerably more recent vintage than the other two; it was introduced by the author [3] in 1978. The crucial fact about A_G that requires checking is that its extremals bear adequate resemblance to the trajectories we are studying, for example the Hamiltonian ones defined by (6). Assuming for simplicity all functions in question to be differentiable, we shall establish the required link <u>under the hypothesis that</u> H(t,x,p) <u>is (jointly) convex in</u> (x,p).

The Euler equation for A_G is

$$d/dt\, [G_q,\ x - G_y] = [\dot{p}, 0],$$

where the partial derivatives of G are of course evaluated at $(t,-\dot{p},\dot{x})$. We rewrite the equation as

$$[x + c,\ p + k] = \nabla G(t,-\dot{p},\dot{x}),$$

where c and k are (unknown) constants. Invoking the inversion formula (9), we derive from this

$$[-\dot{p},\dot{x}] = \nabla H(t, x + c,\ p + k).$$

This simply says that the arc (x + c, p + k) is a Hamiltonian trajectory.

The relation between extremals of the dual action A_G and our trajectories of interest is therefore a bit less direct than it was for the other actions;

we may summarize as follows : extremals of A_G correspond to Hamiltonian trajectories modulo translation by a constant.

2. Existence and necessary conditions

The method of variational principles in its strongest form consists of minimizing an action functional and expressing the necessary conditions for a minimum, thereby obtaining some required type of equation of motion. We shall briefly review in this section the two central ingredients from optimization that figure in this recipe; full details appear in [2].
The paradigm we shall employ is that of the *generalized problem of Bolza*, denoted P_B, which consists of minimizing a functional $J(x)$ of the type

$$J(x) = \ell(x(0),x(T)) + \int_0^T L(t,x,\dot{x})dt$$

where ℓ and L are *extended-valued* functions, and where the class of competing functions is that of all *arcs* (absolutely continuous functions from $\mathbb{R}^n$ to $\mathbb{R}$). The liberty of employing extended-valued functions makes the problem extremely versatile, since various kinds of constraints can be made implicit by defining ℓ and/or L to be $+\infty$ whenever they are violated. We hope to demonstrate in this article the relevance to the variational method of certain new techniques of optimization, namely extended necessary conditions and value functions. We also exploit transversality conditions to a greater extent than seems to have been the case in the past.

2.1. Euler and Hamiltonian inclusions for P_B

Even though ℓ and L are not supposed differentiable, analogues of the classical necessary conditions do exist. These are couched in terms of *generalized gradients* [2]. We shall not recall here the definitions and calculus of generalized gradients; suffice it to say for now that given an extended-valued function f on $\mathbb{R}^n$, and a point x at which f is finite, we may define a set $\partial f(x)$, the generalized gradient of f at x, which reduces to $\{\nabla f(x)\}$ if f is smooth (continuously differentiable), and which in general has many properties similar to the usual derivative.

Suppose now that ℓ is the *indicator function* of the set $\{(A,B)\}$; this means that $\ell(u,v)$ is equal to $+\infty$ unless $u = A$ and $v = B$, in which case it is equal to 0. Then P_B reduces to the minimization of the functional A_L defined by (1), subject to the constraint $x(0) = A, x(T) = B$. Under certain hypotheses which we do not elaborate, a necessary condition for optimality of an arc x in this situation is the existence of another arc p satisfying the "Euler inclusion"

$$(\dot{p}(t), p(t)) \in \partial L(t, x(t), \dot{x}(t)), \tag{14}$$

where ∂L refers to the generalized gradient of L taken with respect to the (x,x) variables. It is easy to see that (14) implies the classical Euler equation (2) when L is smooth and ∂L reduces to $\{\nabla L\}$.

There is also a Hamiltonian formulation of the necessary conditions; now H is defined directly by the formula (10), and the necessary condition asserts the existence of p such that

$$(-\dot{p}(t), \dot{x}(t)) \in \partial H(t, x(t), p(t)), \tag{15}$$

where the generalized gradient is with respect to x and p. Once again this reduces to the classical Hamiltonian system in the presence of smoothness. We refer to [2, Chapter 4] for specifics.

2.2. Transversality conditions

When ℓ is other than the indicator of a point, the Euler or Hamiltonian inclusions above are supplemented by additional conditions involving p(0) and p(T). The general form of these *transversality conditions* is

$$(p(0), -p(T)) \in \partial \ell(x(0), x(T)) \tag{16}$$

These conditions reduce to more familiar ones in typical circumstances. For example, suppose that $\ell(u,v)$ is of the form f(u,v) plus the indicator of a set C in $\mathbb{R}^n \times \mathbb{R}^n$. Then (16) gives :

$$(p(0), -p(T)) \in \partial f(x(0), x(T) + N_c(x(0), x(T)),$$

where N_c is the *normal cone* [2] to C.

2.3. Existence

Space does not permit an extended discussion of the conditions under which P_B admits a solution. We recall only the essence of the classical existence theorem of Tonelli, which requires two conditions on the Lagrangian : convexity and coercivity. The convexity condition is the requirement that for each (t,x), the function $v \to L(t,x,v)$ be convex, while coercivity is the requirement that L(t,x,v) majorize a function of the form $\varepsilon|v|^2 + \beta$ (where $\varepsilon > 0$).

Under these two conditions, the problem of minimizing $A_L(x)$ subject to fixed boundary conditions on x admits a solution. Under a variety of slight additional hypotheses (see [8]) it is possible to assert greater regularity of the solution x to the problem.

3. Periodic solutions via least action

We now seek to apply optimization methods to the three actions defined earlier, with a specific purpose : to attempt to derive the existence of periodic trajectories. In this section we study the possibilities of the simplest approach, that of globally minimizing an action functional. We shall deal only with the autonomous case, in which L (or H, or G) is independent of t.

3.1. The Lagrangian action

In this setting what we seek is a solution x of the Euler equation such that $x(0) = x(T)$ and $\dot{x}(0) = \dot{x}(T)$. We can certainly hope to minimize the Lagrangian action A_L, since the conditions required of L in the Tonelli existence theorem are not unreasonable. This would lead to a solution of the Euler equation. There is a difficulty however with $\dot{x}$, since we seem to have little control over $\dot{x}(0)$ and $\dot{x}(T)$; of course, we also need to know that $\dot{x}$ is continuous.

One thought is to specify a boundary condition directly on $\dot{x}(0)$ and $\dot{x}(T)$, for example, simply that they be equal. This fails to work, however, for the reason that the resulting modified version of P_B will not have a solution in general. This can be seen from the observation that for any arc x feasible for P_B, there exists a further arc y (a modification of x near

0 and T) satisfying $\dot{y}(0) = \dot{y}(T)$ and such that $J(y)$ is as close as desired to $J(x)$.

There are situations in which some control over $\dot{x}(0)$ and $\dot{x}(T)$ can be obtained from the transversality condition. Here is a simple example. We posit the following :

(i) L is C^2 with $L_{vv} > \varepsilon$ (for $\varepsilon > 0$)

(ii) there are n independent elements ζ_i of $\mathbb{R}^n$ such that for each x and v, we have $L(x+\zeta_i, v) = L(x,v)$

PROPOSITION 3.1. Under these hypotheses there exists a periodic solution of the Euler equation.

Proof. It follows easily from (i) that L satisfies the convexity and coercivity conditions. We consider the version of P_B in which $\ell(u,v)$ is the indicator of the set $\{u = v\}$; i.e. we impose $x(0) = x(T)$. Given any feasible arc x, the arc $y = x - \Sigma\lambda_i\zeta_i$ is also feasible for any set of integers $\{\lambda_i\}$ and we have $J(y) = J(x)$ in view of (ii). Since we can always arrange the λ_i to have $|y(0)| \leqslant n \max |\lambda_i|$, it follows that we may redefine $\ell(u,v)$ to equal ∞ when $|u| \geqslant n \max |\lambda_i| + 1$ without essentially changing the problem. With this extra fact, the Tonelli existence theorem applies to guarantee the existence of a solution x to P_B.

We now would like to assert that x satisfies the Euler equation. This would follow from classical results if we knew that $\dot{x}$ is bounded; that this is automatically the case when L is independent of t is a recently proven result of Clarke and Vinter [8, Corollary 3.1]. The fact that x is C^2 also follows [8, Theorem 2.1].

The periodicity will now follow from the transversality condition (16), which in the present situation asserts

$$[p(0), -p(T)] \text{ is normal to } \{(u,v) : u = v\} \text{ at } [x(0), x(T)]$$

This is equivalent to $p(0) = p(T)$. Now the Euler equation implies that $p(t) = L_v(x,\dot{x})$, so we deduce

$$L_v(x(0),\dot{x}(0)) = L_v(x(T),\dot{x}(0)) = L_v(x(T),\dot{x}(T)).$$

Since $v \to L_v$ is one-to-one by (i), it follows that $\dot{x}(0) = \dot{x}(T)$. QED

3.2. The Hamiltonian action

In this context we seek a solution (x,p) of the Hamiltonian system (6) on [0,T] such that $(x,p)(0) = (x,p)(T)$. We immediately perceive an advantage of the Hamiltonian formulation over the Lagrangian, stemming from the fact that we can exercise direct control over all the required boundary values. For example, we could simply impose $(x,p)(0) = (x,p)(T)$ and leave only the differential equation (6) to be verified. Unfortunately the action A_H is very badly behaved from the point of view of optimization.

In seeking to minimize A_H(see (7)) we recognize a serious difficulty, namely that the integrand fails to be coercive. (In fact it is independent of x.) This renders existence of a minimum problematical in general. In fact, it is easy to show that A_H is indefinite (arbitrarily large or small) even if x and p are restricted to bounded sets. For this reason, direct variational methods seem useless in connection with it.

3.3. The dual action

In contrast to A_H, the dual action A_G involves the derivatives of x and p more completely, and there is hope under some circumstances of globally minimizing it. In fact, this was the original motivation for its introduction. We illustrate with a simple example, in which H is convex and subquadratic.

PROPOSITION 3.2. Let H(x,p) be convex, differentiable, nonnegative, and 0 only at the origin. Suppose that H satisfies the following growth condition:

$$a|x|^{1+r} + b|p|^{1+u} \leqslant H(x,p) \leqslant A|x|^{1+R} + B|p|^{1+U} + C$$

where all the constants are positive and r,u,R and U are less than 1. Then for any $T > 0$ there is a periodic nonconstant solution of Hamilton's equations on [0,T] having minimal period T.

Proof. It follows easily from the given growth condition that G, the conjugate of H, satisfies a condition of the form

$$\alpha|\zeta|^{\delta} + \gamma \leqslant G(\zeta) \leqslant \beta|\zeta|^{\sigma} ,$$

where $\alpha,\delta,\beta,\sigma$ are positive constants with δ and σ greater than 2. It then follows from the Tonelli existence theorem that there is a minimum of $A_G(x,p)$ subject to $(x(0),p(0)) = (x(T),p(T)) = (0,0)$. The results of Section 1.3 affirm that a translate (y,q) of the solution (x,p) solves the Hamiltonian equations (note : this is where the convexity of H is invoked). Of course, (y,q) is periodic.

To see that (y,q) is not constant (which would correspond to merely a critical point of H, the origin), it suffices to show that $A_G(y,q)$ $(= \min A_G)$ is strictly negative (for any constant arc gives value 0 to A_G). But in view of the upper estimate for G, it is possible to construct a specific arc for which A_G is negative (see [2, p. 283]), which does the trick.

The final assertion to be proved is that T is the true or minimal period of (y,q); this is a good example of the kind of supplementary information obtainable from a strong variational principle. The proof (see [2, p. 282]) proceeds by demonstrating that if T were not the minimal period of (y,q) then there would be an arc assigning to A_G a value lower than does the solution (x,p), a contradiction. QED

4. Other issues and approaches

One clear lesson to be drawn from the vast and rich literature on periodic solutions is that many different techniques are useful and necessary in the analysis. We have been concentrating upon the most direct variational methods for Hamiltonian systems. In this section we shall briefly discuss more indirect (but still variational) approaches, and also list some references for those wishing to go further.

Much of the recent activity in the area of Hamiltonian systems is due to the rekindling of interest in the subject following the work of Paul Rabinowitz [20] and Alan Weinstein [23]. Recent survey articles worth consulting include those of Desolneux-Moulis [9], Mancini [17], and Rabinowitz [21] . Related references may also be found in [1][12][13][15].

The issue focused upon in the preceding section was that of existence of a solution of prescribed period. Another (the one for which the dual action was first introduced [3]) is the existence of periodic solutions of prescribed energy, an issue treated later in this section. Among the other questions that have been addressed are the number of trajectories, their stability properties, and the topic of forced oscillations. A recurrent theme is the use of critical point theory. To illustrate this approach, we shall follow Ekeland [11] in rederiving a result of Rabinowitz for superquadratic Hamiltonians.

4.1. Superquadratic Hamiltonians

The approach is again based upon the dual action principle, as was Proposition 3.2, but this time the dual action does not admit a global minimum.

PROPOSITION 4.1. Let H satisfy the hypotheses of Proposition 3.2 in addition to being strictly convex, where now we assume that the constants r,u,R and U are greater than 1. Then for any $T > 0$, there is a nonconstant Hamiltonian trajectory of period T.

Proof. We derive the same estimates for G as in the proof of Proposition 3.2, except that now we have δ and σ less than 1. Let us examine the behavior of A_G with this in mind. To begin with, note that the existence of a global minimum is now out of the question, since for large values of the arguments, the indefinite bilinear term $p.x$ will overwhelm the term $G(-\dot{p},\dot{x})$, which behaves like $|(\dot{x},\dot{p})|^d$ for $d < 2$.

On the other hand, for small values of the arguments, the nonnegative term coming from G dominates. In fact, one can show that the arc $(x,p) = (0,0)$ is a local minimum for A_G. Since A_G admits a local minimum at the origin and is indefinite (not bounded below), geometry dictates that A_G admit another critical point. (The precise result used here is the Ambrosetti-Rabinowitz Mountain Pass lemma; the strict convexity of H implies the differentiability of G which the lemma requires). This critical point gives the required Hamiltonian trajectory by the results of Section 1.3.

Note that the minimality of the period no longer follows in this situation, the price of having merely a critical point for A_G rather than a minimum.

QED

4.2. A method of local minima

We shall briefly discuss here the application to periodic trajectories of a recent existence theorem "in the small" of Clarke and Vinter [9]. To describe this result, consider the basic integral functional (1), and its minimization subject to fixed endpoint conditions : $x(0) = A$, $x(T) = B$. Suppose that L is strictly convex in v. In the absence of coercivity, we know that a solution need not exist. However it turns out that if T is sufficiently small, if A is sufficiently close to B, and if the competing arcs are restricted to a certain open set, the resulting modified problem will admit a solution [9]. When applied to A_G, this type of local minimum is adequate to yield an extremal of A_G and hence a Hamiltonian trajectory.

Here is an example of a result derived through this technique (see [7] for details). Note that in contrast to Proposition 3.2 and 4.1, a mix of super- and subquadratic behavior of H is allowed.

PROPOSITION 4.2. Let H be as in Proposition 3.2, except that we now suppose only $ru < 1$, $RU < 1$. Then for every $T > 0$, there is a Hamiltonian trajectory of minimal period T.

4.3. A value function method

We now turn to an example of a new technique that combines value functions and variational principles. The context is that of fixed energy; that is we seek closed Hamiltonian trajectories on a given energy surface $S = \{(x,p) : H(x,p) = h\}$. The hypothesis is made that S is the boundary of a compact convex set containing 0 in its interior. We shall frequently use the notation z for the point (x,p) in $\mathbb{R}^{2n}$.

We define g to be the *gauge* of S; i.e., the function from $\mathbb{R}^{2n}$ to $[0,\infty)$ defined by $g(z)$ = the unique nonnegative λ such that $z \in \lambda S$. It is clear that $S = \{z : g(z) = 1\}$. Thus g also represents the surface S, and in fact there is a correspondence between trajectories on S for H and for g (this fact, due to Rabinowitz, is explained in [2, Proposition 7.7.2]). It suffices therefore to consider the Hamiltonian system for g, which can be written

$$J\dot{z}(t) \in \partial g(z(t)), \qquad (17)$$

where J is the $2n \times 2n$ matrix

$$J = \begin{pmatrix} 0 & -I_n \\ I_n & 0 \end{pmatrix}$$

and I_n is the $n \times n$ identity matrix (g is not differentiable unless the surface S is smooth, so the use of the generalized gradient is made). A curve lying on S which admits a parametrization $z(.)$ on an interval $[0,T]$ satisfying (17) is called a *geodesic* . A $2n \times 2n$ matrix M is called *symplectic* if it satisfies the condition $M^*JM = J$. The identity matrix is symplectic, and provides an interesting special case of the following result [2][6].

PROPOSITION 4.3. If M is symplectic, then for some s in S there is a non-trivial geodesic joining s to Ms.

Proof. When H = g, the dual Hamiltonian G(y,q) reduces to the indicator of a certain convex compact set Σ (the polar of Minkowski dual of S) with the property that for every s,

$$\max \{s.\sigma : \sigma \in \Sigma\} = g(s)$$

We consider a family of parametrized problems P_α, where P_α is the problem consisting of minimizing the functional ϕ defined by

$$\phi(z) = \int_0^1 <J\dot{z},z>dt$$

over the 2n-dimensional arcs z on $[0,1]$ satisfying $J\dot{z}(t) \in \Sigma$ a.e. and the boundary condition $z(1) = Mz(0) + \alpha$. This is a problem which, when reformulated as a generalized problem of Bolza, has Lagrangian equal to $-1/2<J\dot{z},z>$ plus the indicator of Σ. The last term is the conjugate of g, and the first term is relatable to $\dot{p}.x$ when integration by parts is used. These comments indicate that once again we are essentially considering the dual action, this time for the Hamiltonian g.

We let $V(\alpha)$ denote the minimum in the problem P_α. It can be shown that V is finite and Lipschitz near 0 [2, p. 275] . Now it follows from the definition of V that for any arc z we have

$$\phi(z) \geqslant V(z(1) - Mz(0))$$

and that equality holds for any arc y solving P_0. To paraphrase, it follows that y minimizes

$$\phi(z) - V(z(1) - Mz(0))$$

over all arcs z satisfying $Jz \in \Sigma$ a.e. This is a generalized problem of Bolza to which the Hamiltonian necessary conditions of Section 2.1 apply.

To write the necessary conditions (15), we need to calculate the Hamiltonian H^* of the problem; we find without difficulty that $H^*(z,p)$ is given by $g(Jp + z/2)$. Relation (15) is then see to imply

$$J\dot{y} = -2\dot{p} \in \partial g(Jp + y/2).$$

It follows that Jp is a translate of y/2, and that the function $w = Jp + y/2$ satisfies (17).

A solution such as w to (17) is automatically such that $g(w(t))$ is constant : $g(w(t)) = h$. Because $V(0)$ is strictly negative, as is easy to see, it follows that y and hence w is nonconstant, whence $h > 0$.

We now turn to the boundary conditions. The transversality condition (16) in the present situation asserts that for some r, we have $p(1) = r$, $p(0) = M^* r$. Armed with this we calculate

$$\begin{aligned} Mw(0) &= MJp(0) + My(0)/2 \\ &= MJM^* p(1) + y(1)/2 \\ &= Jp(1) + y(1)/2 = w(1) \end{aligned}$$

as required. The only remaining detail is to account for the fact that w lives on the surface $g = h$ rather than $g = 1$ (= S). But because g is positively homogeneous, a simple transformation brings w to S while preserving its required properties. QED

The special case M = I corresponds to a closed orbit. The derivation of its existence was first proved by Rabinowitz [20] and by Weinstein [23]. This case may also be treated by methods which lie entirely within the classical calculus of variations, as shown in [4].

5. Oscillations of the nonlinear pendulum

We now turn to the context of forced oscillations of the nonlinear pendulum, for which the governing equation (in the absence of friction) may be brought to the form

$$x'' + \theta\sin(x) = f(t), \quad 0 \leqslant t \leqslant T, \tag{18}$$

where f lies in $L^1[0,T]$. We are interested in solutions x satisfying $x(0) = x(T)$, $x'(0) = x'(T)$.

A considerable literature exists on this problem; we refer the reader to the excellent survey of Mawhin [18] for details, and to [14][19][24] and the references therein. We remark that the problem of determining all f for which (18) admits a peridic solution remains unsolved.

A variational principle in classical action form is available for the equation (see (4)) ; the appropriate functional is

$$J(x) = \int_0^T \{\frac{|\dot{x}|^2}{2} + \theta \cos(x) + fx\}dt \tag{19}$$

We immediately note some distinctions from the situations considered earlier, the situations in which were derived Propositions 3.2, 4.1 and 4.2. The first of these concerns growth : it is neither sub- nor super-quadratic but rather quadratic. The problems on this dividing line seem to be the most delicate to treat. The second distinction concerns the absence of convexity due to the term cos(x); the straightforward application of the dual action that served us so well above is no longer possible.

We shall proceed to illustrate a method that combines different elements from preceding sections, namely the Lagrangian action (19) with a value function approach similar to that of Section 4.3.

Consider the problem under the added condition

$$\int_0^T f(t)dt = 0 \tag{20}$$

It is not hard then to see that $J(x)$ admits a minimum subject to the boundary condition $x(0) = x(T)$. (In fact, this is essentially a special case of Proposition 3.1.). The minimizer x is then a periodic solution. This approach is due to Willem [24] ,and Proposition 3.1 is a generalization of his result. This global variational method is inapplicable when (20) fails. We now develop an alternate approach which subsumes the global one.

Let us define a family of problems $P(f,\alpha)$ as follows : $P(f,\alpha)$ is the problem of minimizing $J(x)$ subject to the condition $x(0) = x(T) = \alpha$. It is easy to see from the classical Tonelli existence theorem that $P(f,\alpha)$ admits a solution; we denote the minimum value $V(f,\alpha)$, where the notation deliberately keeps track of the forcing term f as well as the parameter α. The basis of our method is that solutions to $P(f,\alpha)$ will be periodic extremals when α is a critical point of V. Here is a precise instance of this assertion :

PROPOSITION 5.1. Let $V(f,.)$ admit a local minimum at a point α. Then any solution x to $P(f,\alpha)$ is a periodic solution of (18).

Proof. We have by assumption $V(f,\alpha) = J(x)$. Let α' be a point near α and let y be feasible for $P(f,\alpha')$. Then we have

$$J(y) \geqslant V(f,\alpha') \geqslant V(f,\alpha) = J(x).$$

The upshot is that x furnishes a local minimum for $J(y)$ subject to the boundary condition $y(0) = y(T)$. Applying the necessary conditions of Section 2.1 yields that x is a smooth extremal (i.e., solution of (18)), and the transversality condition gives $\dot{x}(0) = \dot{x}(T)$. QED

In view of the proposition, we can obtain periodic trajectories by finding conditions that guarantee a local minimum for V. One obvious such condition is (20), for in its presence we have $V(f,\alpha) = V(f,\alpha+2\pi)$ for all α, as is easily seen, which guarantees that $V(f,.)$ has a global minimum. In this light we may view the result of Willem cited above as a special case of the Proposition. And now another avenue is open to us : if f is "close" to f', where f' satisfies (20), then the shape of $V(f,.)$ resembles that of $V(f',.)$, which is periodic. Thus $V(f,.)$, although not necessarily periodic, should admit a local minimum. We shall now give a precise version of this approach when the base function f' is the zero function; thus the condition on f will be that it be "close to 0" in an appropriate sense. In any case, f certainly cannot be too large, since a necessary condition for (18) to admit a periodic solution is that

$$\left|\int_0^T f(t)dt\right| \leqslant \theta T$$

We begin with a technical lemma concerning $V(0,.)$.

LEMMA 1. $V(0,0) \geqslant \theta T\{1-\theta T^2/24\}$

In the minimum defining $V(0,.)$, it makes no difference (in view of symmetry) to require $x(t) \geqslant 0$. For $x \geqslant 0$, we have $\cos(x) \geqslant 1-x$, so that $V(0,0)$ is bounded below by the minimum of

$$\int_0^T \{\frac{|\dot{x}|^2}{2} + \theta(1-x)\}dt$$

subject to $x(0) = x(T) = 0$. This last minimum can be explicitly evaluated by elementary calculus of variations, and is found to be the right side in the statement of the lemma.

LEMMA 2. Let x solve $P(f,0)$. Then

$$|fx|_1 \leqslant T|f|_1\{\theta T + 4|f|_1\}/\theta,$$

where $|.|_1$ denotes the L^1 norm on $[0,T]$.

We know that x satisfies (18). Let τ be a point in $[0,T]$ at which $|x(t)|$ assumes its maximum, so that $x'(\tau) = 0$. Assume for now that τ is no greater than T/2, and consider s in $[0,\tau]$. From the relation

$$x'(s) = x'(\tau) + \int_s^\tau x''$$

together with (18) we derive

$$|x'(s)| \leqslant \theta|s-\tau| + |f|_1$$

We use this last relation to deduce

$$|x(\tau)| = |\int_0^\tau x'| \leqslant \theta\tau^2/2 + \tau|f|_1$$

Since $|x(t)|$ assumes its maximum at $t = \tau$ and τ lies in $[0,T/2]$, the inequality of the lemma now results immediately. The same relation follows in the case $\tau > T/2$ by integrating in $[\tau, T]$ rather than $[0,T]$.

We now turn to the main result of the section.

PROPOSITION 5.2. Let γ be the solution in $[\pi/2,\pi]$ of the equation

$$T\theta \sin(\gamma) = |\int_0^T f(t)dt|$$

and suppose that f satisfies

$$\theta\cos(\gamma) + \gamma|\int_0^T f(t)dt|/T + |f|_1\{\theta T + 4|f|_1\}/8 \leqslant \theta\{1 - \theta T^2/24\}$$

Then there is a periodic solution of (18).

Proof. In view of Proposition 5.1, we need only prove that $V(f,.)$ admits a minimum on the open interval $(0,2\pi)$. Since $V(f,.)$ is continuous, it suffices to establish that $\min\{V(f,\alpha) : 0 \leqslant \alpha \leqslant 2\pi\}$ is strictly less than both $V(f,0)$ and $V(f,2\pi)$. Let us deal with the case in which f has non-negative integral over $[0,T]$. Then $V(f,2\pi) \geqslant V(f,0)$, so we seek to establish

$$\min\{V(f,\alpha) : 0 \leqslant \alpha \leqslant 2\pi\} < V(f,0) \tag{21}$$

It is certainly the case that $V(f,\alpha)$ is bounded above by $J(z)$, where z is the arc identically equal to α. We find that $J(z)$ equals

$$T\theta \cos(\alpha) + \alpha \int_0^T f(t)dt$$

One of the possible α is γ itself, whence the left side of (21) is bounded above by

$$T\theta \cos(\gamma) + \gamma\int_0^T f(t)dt \qquad (22)$$

Now let x solve $P(f,\alpha)$. Then invoking Lemmas 1 and 2 above we derive

$$\begin{aligned} V(f,0) &= J(x) \\ &\geqslant \int_0^T \{\frac{|\dot{x}|^2}{2} + \theta \cos(x)\}dt - |\int_0^T fx| \\ &\geqslant V(0,0) - T|f|_1\{\theta T + 4|f|_1\}/8 \\ &\geqslant \theta T\{1 - \theta T^2/24\} - T|f|_1\{\theta T + 4|f|_1\}/8 \end{aligned}$$

The inequality in the statement of the proposition implies that this last quantity is greater than (22), and (21) follows as required. QED

References

[1] J.P. Aubin and I. Ekeland, Applied Nonlinear Analysis, John Wiley and Sons, New York (1984).

[2] F.H. Clarke, Optimization and Nonsmooth Analysis, John Wiley and Sons, New York (1983).

[3] F.H. Clarke, Solutions périodiques des équations hamiltoniennes, Comptes Rendus Acad. Sci. Paris 287 (1978), pp. 951-952.

[4] F.H. Clarke, A classical variational principle for periodic Hamiltonian trajectories, Proc. Amer. Math. Soc. 76 (1979), pp. 186-188.

[5] F.H. Clarke, Periodic solutions of Hamiltonian inclusions, J. Differential Equations 40 (1981), pp. 1-6.

[6] F.H. Clarke, On Hamiltonian flows and symplectic transformations, SIAM J. Control Optim. 20 (1982), pp. 355-359.

[7] F.H. Clarke, Periodic trajectories and local minima of the dual action, Trans. Amer. Math. Soc. 287 (1985), pp. 239-251.

[8] F.H. Clarke and R.B. Vinter, Regularity properties of solutions to the basic problem in the calculus of variations, Trans. Amer. Math. Soc. 289 (1985), pp. 73-98.

[9] F.H. Clarke and R.B. Vinter, Regularity and existence in the small in the calculus of variations, J. Differential Equations, in press.

[10] N. Desolneux-Moulis, Orbites périodiques des systèmes hamiltoniens autonomes, Séminaire Bourbaki 32 (1979), No. 552.

[11] I. Ekeland, Periodic Hamiltonian trajectories and a theorem of Rabinowitz, J. Differential Equations 34 (1979), pp. 523-534.

[12] I. Ekeland and H. Hofer, Periodic solutions with prescribed period for convex autonomous Hamiltonian systems, to appear.

[13] I. Ekeland and J.M. Lasry, On the number of periodic trajectories for a Hamiltonian flow on a convex energy surface, Ann. of Math. 112 (1980), pp. 283-319.

[14] G. Fournier and J. Mawhin, On periodic solutions of forced pendulum-like equations, J. Differential Equations, to appear.

[15] M. Girardi and M. Matzeu, Some results on solutions of minimal period to Hamiltonian systems, in Nonlinear Oscillations for Conservative Systems (Venice (1985); A. Ambrosetti, Ed.), Pitagora Editrice, Bologna .

[16] H.H. Goldstine, A History of the Calculus of Variations, Springer-Verlag, New York (1980).

[17] H. Mancini, Periodic solutions of Hamiltonian systems having prescribed minimal period, in Advances in Hamiltonian Systems, Birkhauser, Boston, Mass. (1983).

[18] J. Mawhin, Periodic oscillations of forced pendulum-like equations, in Ord. and Partial Diff. Equations (Dundee (1982); Everitt and Sleeman, Eds.), Lecture Notes in Math. 964 (1982), Springer, Berlin pp. 458-476.

[19] J. Mawhin and M. Willem, Multiple solutions of the periodic boundary-value problem for some forced pendulum-like equations, J. Differential Equations 52 (1984), pp. 264-287.

[20] P.H. Rabinowitz, Periodic solutions of Hamiltonian systems, Comm. Pure and Appl. Math. 31 (1978) pp. 157-184.

[21] P.H. Rabinowitz, Periodic solutions of Hamiltonian systems : a survey, SIAM J. Math. Anal. 13 (1982), pp. 343-352.

[22] R.T. Rockafellar, Convex Analysis, Princeton University Press, Princeton, N.J. (1970).

[23] A. Weinstein, Periodic orbits for convex Hamiltonian systems, Ann. of Math. (2) 108 (1978), pp. 507-518.

[24] M. Willem, Oscillations forcées de systèmes hamiltoniens, Publications de l'Université de Besançon (1981).

Frank H. Clarke
Centre de Recherches Mathématiques
Université de Montréal
Case postale 6128, succursale A
Montréal (Quebec) H3C 3J7
Canada

A HARAUX & V KOMORNIK

Oscillations in the wave equation

1. Introduction

A typical problem which motivated the investigation of the present paper, is the following : considering all the possible vibrations of a membrane with fixed boundary, how long can a given point of the membrane remain strictly above the equilibrium position ?

To be more specific, let Ω be a bounded domain in $\mathbb{R}^N$, $N \geqslant 1$ and set *

$U = \{u \in C^{\infty}(\mathbb{R} \times \bar{\Omega}),\ u''-\Delta u = 0 \text{ in } \mathbb{R} \times \Omega,\ u = 0 \text{ on } \mathbb{R} \times \partial\Omega\}$

$T(x,u) = \sup \{t>0 \mid u(s,x)>0,\ \forall s \in [0,t]\}$ for any $x \in \Omega$ and $u \in U$.

If $N = 1$ and $\Omega =]0,\ell[$ then $T(x,u) \leqslant 2\ell$, $\forall x \in \Omega$, $\forall u \in U$, because $u(.,x)$ is 2ℓ-periodic with mean-value 0.

If $N > 1$, then it follows from a result of T. Cazenave and A. Haraux [3] that $T(x,u) < +\infty$ for all $x \in \Omega$ and $u \in U$.

We will show in this paper that in some cases $\sup\limits_{u \in U} T(x,u) = +\infty$ for "almost all $x \in \Omega$" if $N > 1$.

2. Positive orthogonal functions

Let $0 < T < +\infty$, $X \subset C([0,T])$, $h \in C([0,T])$ and suppose that

$$h > 0 \qquad \text{on }]0,T[\,, \tag{1}$$

$$\int_0^T h\varphi = 0 \quad \forall \varphi \in X. \tag{2}$$

Then obviously any function $\varphi \in X$ which is not identically 0, must change sign in $]0,T[$. This simple remark shows the role played by positive orthogonal functions in oscillation properties.

---

* All functions in this paper are supposed to be real-valued.

Now let $\tau > 0$ be arbitrary : we define

$$X_\tau = \{f \in C(\mathbb{R}) \mid f \text{ is } \tau\text{-periodic and } \int_0^\tau f = 0\}. \tag{3}$$

The main result of this section is the following

THEOREM 1. Let $\tau_1, \ldots, \tau_n$ be a finite sequence of positive numbers and put $T := \sum_1^n \tau_j$. Then there exists a function $h \in C([0,T])$ such that

$$h(t) > 0 \qquad \forall t \in]0,T[\;, \tag{4}$$

$$\int_0^T h\varphi = 0 \qquad \forall \varphi \in \sum_1^n X_{\tau_j}. \tag{5}$$

Proof. We define the functions $h_1, \ldots, h_n : \mathbb{R} \to \mathbb{R}$ and $h : [0,T] \to \mathbb{R}$ as follows:

$$h_1(t) = \begin{cases} 1 & \text{if } t \in [0, \tau_1], \\ 0 & \text{if } t \in \mathbb{R} \setminus [0, \tau_1], \end{cases} \tag{6}$$

$$h_k(t) := \int_0^{\tau_k} h_{k-1}(t-s)\,ds, \; \forall t \in \mathbb{R}, \; k = 2, \ldots, n, \tag{7}$$

$$h := h_n \big|_{[0,T]}. \tag{8}$$

(Constructions of this type were introduced for other purposes by S. Mandelbrojt.)

One can easily see that $h \in C([0,T])$, $h > 0$ on $]0,T[$ and supp $h = [0,T]$. To finish the proof we show by induction on $k = 1, \ldots, n$ that

$$\int_{\mathbb{R}} h_k\varphi = 0 \qquad \forall \varphi \in \bigcup_1^k X_{\tau_j}. \tag{9}$$

For $k = 1$ this follows immediately from definitions (3) and (6). Suppose now that $2 \leq k \leq n$ and (9) is true for $k-1$ instead of k. For any

$\varphi \in \bigcup_{1}^{k} X_{\tau_j}$ we have

$$\int_{\mathbb{R}} h_k \varphi = \int_{R} \int_{0}^{\tau_k} h_{k-1}(t-s)ds\, \varphi(t)dt =$$

$$= \int_{0}^{\tau_k} \int_{\mathbb{R}} h_{k-1}(t-s)\varphi(t)dt\; ds = \int_{0}^{\tau_k} \int_{\mathbb{R}} h_{k-1}(u)\varphi(u+s)du\; ds. \tag{10}$$

If $\varphi \in \bigcup_{1}^{k-1} X_{\tau_j}$ then

$$\int_{\mathbb{R}} h_{k-1}(u)\; \varphi(u+s)du = 0 \quad \forall s \in \mathbb{R}$$

by the inductive hypothesis.

If $\varphi \in X_{\tau_k}$ then

$$\int_{0}^{\tau_k} \int_{\mathbb{R}} h_{k-1}(u)\varphi(u+s)du\, ds = \int_{\mathbb{R}} h_{k-1}(u) \int_{0}^{\tau_k} \varphi(u+s)ds\; du$$

$$= \int_{\mathbb{R}} h_{k-1}(u).0\; du = 0 \tag{12}$$

by the definition of X_{τ_k}.

Finally, (9) follows from (10), (11) and (12). ■

REMARK. In the case when n = 2 and $\tau_1 > \tau_2$, the function h has the following graph :

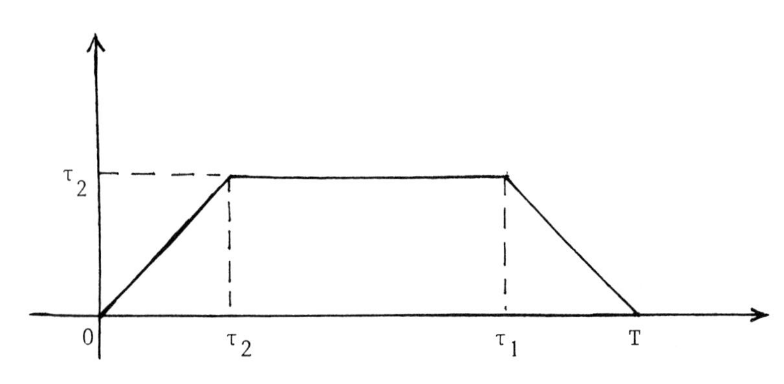

The following infinite version of Theorem 1 is also valid :

THEOREM 2. Let $(\tau_j)_{j=1}^{\infty}$ be a sequence of positive numbers such that $T := \sum_1^{\infty} \tau_j < +\infty$. Then there exists a function $h \in C([0,T])$ such that

$$h(t) > 0 \qquad \forall t \in]0,T[, \tag{13}$$

$$\int_0^T h\varphi = 0 \qquad \forall \varphi \in \bigcup_1^{\infty} X_{\tau_j} . \tag{14}$$

Sketch of the proof. We define the functions $h_k : \mathbb{R} \to \mathbb{R}$ by the formulas (6), (7) for $k = 1,2,\ldots$ and we show that after normalizing in $L^1(\mathbb{R})$ the resulting sequence converges in the space $C_b(\mathbb{R})$ to a function h_∞. Then we show that the function $h := h_\infty|_{[0,T]}$ satisfies the conditions (13), (14). For the details see [5]. □

3. Density theorems

The following result shows that Theorem 1 is in a sense optimal :

THEOREM 3. Let $\tau_1,\ldots,\tau_n$ be a finite sequence of positive numbers such that

$$\frac{\tau_i}{\tau_j} \notin Q \quad \text{if } i \neq j . \tag{15}$$

Then $\sum_1^n X_{\tau_j}$ is dense in $C([0,T])$ for all $T \in]0, \sum_1^n \tau_j[$. ■

Theorem 3 is a consequence of a more general result in the theory of mean-periodic functions, see [8]. We shall now prove a stronger result which shows that the above result is "almost" true for $T = \sum_1^n \tau_i$. Let us denote by $\hat{X}_\tau$ the space of τ-periodic trigonometric polynomials with mean 0 (this is a dense subset of X_τ with respect to the topology of uniform convergence) and by P_k the space of algebraic polynomials of degree $\leq k$.

THEOREM 4. Let $\tau_1,\ldots,\tau_n$ be a finite sequence of positive numbers satisfying (15) and $T:=\sum_1^n \tau_j$. Then $P_n + \sum_1^n \hat{X}_{\tau_j}$ is dense in $C([0,T])$.

For the proof we shall need two simple lemmas (they can be verified by direct calculation or see [5], [6]) :

LEMMA 1. $\forall\, \tilde{p}\in P_{n-1}\quad \exists\, p\in P_n$ such that

$$p(t+\tau_n) - p(t) = \tilde{p}(t) \quad \forall t\in \mathbb{R}. \qquad \square$$

LEMMA 2. Let σ and τ be two positive numbers with $\frac{\sigma}{\tau} \notin Q$. Then $\forall \tilde{f}\in \hat{X}_\sigma \quad \exists\, f\in \hat{X}_\sigma$

$$f(t+\tau_n) - f(t) = \tilde{f}(t) \quad \forall t\in \mathbb{R} . \qquad \square$$

Proof of Theorem 4. By induction on n.

For n = 1 the assertion is true because $P_1 + X_{\tau_1} = C([0,\tau_1])$ and $\hat{X}_{\tau_1}$ is dense in X_{τ_1}.

Suppose now that $n\geqslant 2$ and the theorem is true for n-1. Let $f\in C([0,T])$ and $\varepsilon > 0$ be given arbitrarily.

First step. We define $\tilde{f}\in C([0,T-\tau_n])$ by the formula

$$\tilde{f}(t) = f(t+\tau_n) - f(t) \quad \forall t\in [0,T-\tau_n]. \tag{16}$$

Applying the inductive hypothesis, there exist $\tilde{p}\in P_{n-1}$ and $\tilde{f}_j\in \hat{X}_{\tau_j}$, $j = 1,\ldots,n-1$, such that

$$\|\tilde{f} - \tilde{p} - \sum_1^{n-1} \tilde{f}_j\|_{C([0,T-\tau_n])} < \varepsilon. \tag{17}$$

Second step. Using the lemmas, there exist $p\in P_n$ and $f_j\in \hat{X}_{\tau_j}$, $j = 1,\ldots,n-1$, such that

$$\tilde{p}(t) = p(t+\tau_n) - p(t) \quad \forall t\in \mathbb{R}, \tag{18}$$

$$\tilde{f}_j(t) = f_j(t+\tau_n) - f_j(t) \quad \forall t \in \mathbb{R},\ j = 1,\dots,n-1 \tag{19}$$

Third step. Applying the case n = 1 of the present theorem to the function

$$f - p - \sum_1^{n-1} f_j \in C([0,\tau_n]),$$

there exist $q \in P_1$ and $f_n \in X_{\tau_n}$ with

$$\| (f - p - \sum_1^{n-1} f_j) - q - f_n \|_{C([0,\tau_n])} < \varepsilon. \tag{20}$$

Fourth step. It follows from (16) - (20) that

$$\| (f - p - \sum_1^{n-1} f_j) - q - f_n \|_{C([0,T])} < (1 + \frac{4T}{\tau_n}).\varepsilon. \tag{21}$$

Indeed, putting

$$g = (f - p - \sum_1^{n-1} f_1) - q - f_n,$$

in view of (20) it suffices to show that

$$|g(t+\tau_n) - g(t)| < 4\varepsilon \quad \forall t \in [0, T - \tau_n]. \tag{22}$$

To see this, first we observe that

$$|g(t+\tau_n) - g(t)| = |(\tilde{f} - \tilde{p} - \sum_1^{n-1} \tilde{f}_j)(t) - q(t+\tau_n) + q(t)| \tag{23}$$

$$< \varepsilon + |q(t+\tau_n) - q(t)|$$

by (16), (17), (18), (19) and $f_n \in X_{\tau_n}$.

Furthermore, it follows from the same properties and from (20) that

$$|q(\tau_n) - q(0)| < 3\varepsilon. \tag{24}$$

Having $q \in P_1$, $q(t+\tau_n) - q(t) = q(\tau_n) - q(0)$ and (22) follows from (23), (24). Having $\varepsilon > 0$ arbitrary, from (21) the theorem follows because $p + q \in P_n$ and $f_j \in \hat{X}_{\tau_j}$, $j = 1,\ldots,n$. ■

Remark. The theorems 3, 4 can be generalized for $C^m([0,T])$ instead of $C([0,T])$, $m = 0,1,2,\ldots$; see [5], [6].

4. Applications to the problems of oscillation

The first result of this section solves the problem raised in the introduction.

THEOREM 5. Let a and b be two positive numbers such that

$$\frac{b^2}{a^2} \notin Q, \tag{25}$$

$\Omega :=]0,a[\times]0,b[$ and fix $(x_0,y_0) \in \Omega$ with

$$\frac{x_0}{a} \notin Q \quad \text{and} \quad \frac{y_0}{b} \notin Q. \tag{26}$$

Then to each $T \in]0,+\infty[$ there exists a function u such that

$$u \in C^\infty(\mathbb{R} \times \bar{\Omega}),\ u'' - \Delta u = 0 \text{ in } \mathbb{R} \times \Omega,\ u = 0 \text{ over } \mathbb{R} \times \partial\Omega . \tag{27}$$

$$u(t, x_0, y_0) > 1 \qquad \forall t \in [0,T]. \tag{28}$$

Proof. We put

$$\tau_n := \frac{2ab}{\sqrt{b^2n^2 + a^2}} \quad , \ n = 1,2,\ldots \tag{29}$$

and we fix a natural number N such that

$$\sum_1^N \tau_n > T. \tag{30}$$

For any *finite* double sequences $(u_{m,n})_{m=1,n=1}^{M \quad N}$, $(\alpha_{m,n})_{m=1,n=1}^{M \quad N}$ of real numbers ($M \in \mathbb{N}$ arbitrary), the function

$$u(t,x,y) \equiv$$

$$\sum_{n=1}^{N} \sum_{m=1}^{M} u_{m,n} \cdot \cos\left(\frac{m\pi}{ab}\sqrt{b^2n^2+a^2}\,t+\alpha_{m,n}\right) \cdot \frac{\sin \frac{mn\pi x}{a}}{\sin \frac{mn\pi x_0}{a}} \cdot \frac{\sin \frac{m\pi y}{b}}{\sin \frac{m\pi y_0}{b}}$$

(which is well-defined by the condition (26)) obviously has the properties (27). Furthermore, the function

$$u(t,x_0,y_0) \equiv \sum_{n=1}^{N} \sum_{m=1}^{M} u_{m,n} \cdot \cos\left(\frac{m\pi}{ab}\sqrt{b^2n^2+a^2}\,t + \alpha_{m,n}\right)$$

is an arbitrary element of the space $\sum_{n=1}^{N} \hat{X}_{\tau_n}$ (which is dense in $\sum_{n=1}^{N} X_{\tau_n}$ with the topology of uniform convergence). The condition (15) of Theorem 3 is satisfied by (25) and (29). In view of (30) we can therefore apply Theorem 3 and the theorem follows : we can choose the numbers $u_{m,n}$, $\alpha_{m,n}$ so as to have the property (28). ■

Remark. Modifying the above proof, Y. Meyer proved [9] that Theorem 5 remains valid if we replace the condition (25) by the condition $a = b$. ■
The following theorem is a positive result of oscillation.

Theorem 6. Let Ω be a bounded domain in $\mathbb{R}^2$ with smooth boundary (of class C^∞). Then there exists a positive number T such that for any function u such that

$$\begin{cases} u \in C^\infty(\mathbb{R} \times \bar{\Omega}), \quad u'' - \Delta^3 u = 0 \text{ in } \mathbb{R} \times \Omega, \\ u = \Delta u = \Delta^2 u = 0 \text{ on } \mathbb{R} \times \partial\Omega \end{cases} \tag{31}$$

and for any $x \in \Omega$,
either $u(t,x) = 0 \quad \forall t \in [0,T]$,

or $\exists\, t_1, t_2 \in [0,T]$ with $u(t_1,x) < 0$ and $u(t_2,x) > 0$.

Sketch of the proof. Let us denote by (λ_n) the eigenvalues (with multiplicity) of the operator $-\Delta^3$ with the boundary conditions

$u = \Delta u = \Delta^2 u = 0$ on $\partial\Omega$, and put $\tau_n = \frac{2\pi}{\sqrt{\lambda_n}}$, $n = 1,2,\dots$. We can show that

for any function u satisfying (31) and for any $x \in \Omega$,

$$u(.,x) \in \overline{\mathrm{Vect}}\left(\bigcup_1^\infty X_{\tau_n}\right)$$

(closure in $C_b(\mathbb{R})$). On the other hand, one can check that $\sum_1^\infty \tau_n < +\infty$, and the theorem follows from Theorem 2. For the details and also for more general results see [5]. ■

5. Characterization of the orthogonal functions

It follows from the result stated below that in the case when the periods τ_j are pairwise uncommensurable, the function h constructed in the proof of Theorem 1 is (apart from a multiplicative constant) the unique function satisfying (4) and (5).

THEOREM 7. Let $\tau_1,\dots,\tau_n$ be a finite sequence of positive numbers satisfying (15), $T := \sum_1^n \tau_j$ and set

$$H := \{h \in C([0,T])' \mid \langle h,\varphi\rangle = 0 \quad \forall \varphi \in \sum_1^n X_{\tau_j}\}.$$

($C([0,T])'$ denotes the dual of $C([0,T])$.). Then H is finite-dimensional. Moreover, the set $\{h_n, h_n', \dots, h_n^{(n)}\}$ where h_n is defined by (6),(7) and the derivatives are taken in distributional sense, is a basis for H. ■

The proof of this result is not difficult by applying Theorem 4. A more general result is proved in [6].

Bibliography

[1] L. Amerio, G. Prouse, Almost periodic functions and functional equations. Van Nostrand, New York (1971).

[2] T. Cazenave, A. Haraux, Propriétés oscillatoires des solutions de certaines équations des ondes semi-linéaires. C.R.A.S. Paris, Série I. t. 298 (1984), pp. 449-452.

[3] T. Cazenave, A. Haraux, On the nature of free oscillations associated with some semi linear wave equations, in "Research Notes in Mathematics" no. 122, H. Brezis & J.L. Lions Editors, Nonlinear Partial Differential Equations and their Applications, Collège de France Seminar, Pitman (1984-85), pp. 59-79.

[4] A. Haraux, Nonlinear Evolution Equations - Global Behavior of Solutions. Lecture Notes in Mathematics no. 841, Springer (1981).

[5] A. Haraux, V. Komornik, Oscillations of anharmonic Fourier series and the wave equation, Revista Matemática Iberoamericana 1, 4 (1985), pp. 57-77.

[6] V.Komornik, Oscillatory problems and orthogonal distributions. To appear.

[7] Y. Meyer, Nombres premiers et vibrations. Séminaire Delange-Pisot-Poitou (1972).

[8] Y. Meyer, Algebraic numbers and harmonic analysis, North-Holland publish. Co. (1970).

[9] Y. Meyer, Personal communication.

A. Haraux
Analyse Numérique,
Tour 55-65, 5e étage
Université Pierre et Marie Curie
4, place Jussieu
75230 Paris Cedex 05
France

V. Komornik
Eötvös Loránd University
Department of Analysis
H-1088 Budapest
Múzeum Krt. 6-8
Hungary

D JERISON

Unique continuation and Carleman-type inequalities

In this note we will discuss some Carleman-type inequalities that lead to a unique continuation theorem for Schrödinger operators. The unique continuation theorem shows in turn that certain Schrödinger operators have no positive eigenvalues. The results presented here are joint work with Carlos E.Kenig. Full proofs will appear in a forthcoming article [9].

Consider the Laplace operator $\Delta = -\sum_{j=1}^{n} \partial^2/\partial x_j^2$ on $\mathbb{R}^n$. Denote by $H^{k,p}(\mathbb{R}^n)$ the space of functions f such that $\partial^\alpha f/\partial x^\alpha$ belongs to $L^p(\mathbb{R}^n,dx)$ for $|\alpha| \leqslant k$; for an open subset Ω of $\mathbb{R}^n$, denote by $H^{k,p}(\Omega,\mathrm{loc})$ the space of functions f on Ω such that φf belongs to $H^{k,p}(\mathbb{R}^n)$ for every smooth function φ of compact support in Ω.

THEOREM 1. Suppose that $n \geqslant 3$, $V \in L^{n/2}(\mathbb{R}^n,dx)$, and V has compact support. If $u \in H^{1,2}(\mathbb{R}^n)$ is a weak solution of $\Delta u + Vu = \lambda u$ and λ is a positive constant, then u is identically zero. ■

This theorem says that the Schrödinger operator $\Delta + V$ has no positive eigenvalues. The proof goes as follows. We observe that $H^{1,2}(\mathbb{R}^n) \subset L^p(\mathbb{R}^n,dx)$ for $1/p = 1/2 - 1/n$, in other words, $p = 2n/(n-2)$. Hence $Vu \in L^q(\mathbb{R}^n,dx)$ for $1/q = 1/p + 2/n$, in other words $q = 2n/(n+2)$. Standard regularity estimates for the Laplace operator now imply that $u \in H^{2,q}(\mathbb{R}^n,\mathrm{loc})$. Moreover a well-known argument using the expansion of u in a neighborhood of infinity in terms of spherical harmonics shows that u vanishes identically in a neighborhood of infinity. (Indeed, much weaker conditions on V than compact support yield this conclusion. See, for example, [1,5] .) In particular, u vanishes in an open set and satisfies the differential inequality $|\Delta u| \leqslant |(\lambda - V)u|$. Theorem 1 now follows from

THEOREM 2. Suppose that Ω is a connected open subset of $\mathbb{R}^n$, $n \geqslant 3$. Suppose that $a \in L^{n/2}(\Omega,dx)$, $u \in H^{2,q}(\Omega,\mathrm{loc})$ with $q = 2n/(n+2)$ and $|\Delta u(x)| \leqslant |a(x)u(x)|$ for almost every $x \in \Omega$. If for some $x_0 \in \Omega$ and every $N \geqslant 0$

$$\int_{|x-x_0|<\varepsilon} |u(x)|^2 dx = O(\varepsilon^N) \text{ as } \varepsilon \to 0, \text{ then } u \text{ is identically zero on } \Omega.$$ ■

Theorem 2 is called a "strong" unique continuation theorem because we have only assumed that u vanishes to infinite order at a single point x_0. In order to prove Theorem 1 it would have sufficed to prove an ordinary unique continuation theorem in which one supposes that u vanishes in an open set. The first such unique continuation theorem was proved by Carleman in $\mathbb{R}^2$ with bounded potential a(x). All subsequent work follows from his basic idea of Carleman-type inequalities.
The technique has also proved useful for other uniqueness questions. (See [7] Chapter 8, where further references may be found). Unbounded potentials a(x) were considered recently by many authors. (See Georgescu [6], Schechter and Simon [12], Amrein, Berthier and Georgescu [2], Saut and Scheurer [10] and the survey of Simon [13] for further references.) The best results preceding our Theorem 2 require $a \in L^w$, $w > n/2$ for $n = 2,3,4$ (Amrein, and all [2]) and $a \in L^w$, $w \geqslant (4n-2)/7$ for $n > 4$ (Hörmander [8]). The following example shows that the exponent n/2 is sharp :

Let $\varepsilon > 0$ and $u(x) = \exp(-(\log 1/|x|)^{1+\varepsilon})$ for $|x| < 1$. Then u vanishes to infinite order at 0 and $\Delta u(x) = a(x)u(x)$ for $a(x) \simeq (\log \frac{1}{|x|})^{2\varepsilon} / |x|^2$.
Thus $a(x) \in L^w(R^n, \text{loc})$ for all $w < n/2$.

This examples leaves the possibility that it suffices for a(x) to be of weak type $L^{n/2}$. Indeed, E.M. Stein has simplified and extended our result to the case of coefficients a(x) in weak $L^{n/2}$ with small norm. We will describe this simplification at the end of the note. It is also possible to ask for unique continuation theorems involving different function spaces from L^w classes proposed by B. Simon [13]. E. Sawyer has recently obtained a result of this kind when n = 3 [11].

The Carleman-type inequality we will need is the following.

<u>THEOREM 3</u>. Let n be an integer $\geqslant 3$, $p = 2n/(n-2)$ and $q = 2n/(n+2)$. Suppose that $t \in \mathbb{R}$ is not an integer and δ is the distance from t to the nearest integer. There is a constant C depending only on δ and n such that

$$\| \, |x|^{-t} f \|_{L^p(\mathbb{R}^n, \frac{dx}{|x|^n})} \leqslant C \, \| \, |x|^{-t+2} \, \Delta f \, \|_{L^q(\mathbb{R}^n, \frac{dx}{|x|^n})}, \tag{*}$$

for every $f \in C_0^\infty(\mathbb{R}^n \setminus \{0\})$. ■

In general Carleman's method says that an inequality of the form (*) for a sequence of values of t tending to $+\infty$ yields a unique continuation theorem for $u \in H^{2,q}$ and $a \in L^w$ for $1/q - 1/p = 1/w$. (See [2], [9]) What is crucial is that the constant C be independent of the value of t in the sequence.

Several comments are in order to explain the nature of the inequality (*). First, we have assumed that f vanishes in a neighborhood of the origin in order to make sense out of the norms when the weight $|x|^{-t}$ is very singular. (The change of variables by inversion $x \to x/|x|^2$ shows that the inequality for t is equivalent to the inequality for (n-2)-t.) Second, the spaces $L^p(\mathbb{R}^n, dx/|x|^n)$ are dilation invariant, that is, $f(x) \to f(\delta x)$ preserves the norm. Thus any inequality of the form (*) for any value of p,q and t is invariant under dilation. However, if we consider functions supported in $1 < |x| < 2$ we see from standard regularity estimates for the Laplacian that necessarily $1/q - 1/p \leq 2/n$. Finally, for any fixed non-integer value of t, the inequality (*) with the limiting case $1/q - 1/p = 2/n$ is not hard to prove. What is difficult (and, as emphasized above, essential to Carleman's method) is to prove that the constant does not depend on t.

A useful observation that led us to the proof of Theorem 3 is that for the special values of $q = 2n/(n+2)$ and $w = n/2$, the exponents p and q are dual : $1/p + 1/q = 1$. This recalls the classical Hausdorff-Young inequality that the Fourier transform is bounded from L^q to L^p whenever p and q are dual exponents and $q \leq p$. The Hausdorff-Young inequality is proved by interpolation of an $L^2 \to L^2$ bound and a (trivial) $L^1 \to L^\infty$ bound. The interpolation used to prove Theorem 3 differs from the classical case in that not only the function space, but also the operator varies with a complex parameter. This analytic interpolation procedure was developed by E.M Stein ([15] p.205). The advantage of splitting the estimate into two parts is that in the $L^2 \to L^2$ case we can use spherical harmonics. Most of the earlier approaches to this problem involved spherical harmonics directly in the $L^q \to L^p$ case. As a result the orthogonality of the spherical harmonics was not exploited to its fullest extent. On the other hand, we need an entirely different argument in the $L^1 \to L^\infty$ case. In the

$L^2 \to L^2$ case we look at the symbol of the operator and in the $L^1 \to L^\infty$ case we look at its kernel.

In order to carry out the interpolation we first interpret (*) as an operator inequality. For $f \in C_0^\infty(\mathbb{R}^N \setminus \{0\})$, denote $g(x) = |x|^{-t+2}\Delta f(x)$. Then (*) can be rewritten as

$$\| \, |x|^{-t} \, \Delta^{-1}(|x|^{t-2}g) \|_{L^p(\mathbb{R}^n, \frac{dx}{|x|^n})} \leqslant C \, \|g\|_{L^q(\mathbb{R}^n, \frac{dx}{|x|^n})} \, . \qquad (*')$$

Roughly speaking we will introduce the family of operators "$T_z g = |x|^{-t} \, \Delta^{-z/2}(|x|^{t-z}g)$".

In order to make this precise, recall that for $h \in C_0^\infty(\mathbb{R}^n)$, $0 < \mathrm{Re}\, z < n$,

$$\Delta^{-z/2}h(x) = \int_{\mathbb{R}^n} c_z|x-y|^{z-n}h(y)\,dy$$

with

$$c_z = \pi^{-n/2}2^{-z}\Gamma(\tfrac{n}{2} - \tfrac{z}{2}) \, / \, \Gamma(z/2); \text{ see [14].}$$

If we define $T_2 g(x) = |x|^{-t}f(x)$, then with the notation above for $g(x)$,

$$T_2 g(x) = |x|^{-t} \, \Delta^{-1}(|x|^{t-2}g(x)) = |x|^{-t}\int c_2|x-y|^{2-n}g(y)|y|^{t-2}dy.$$

Because f vanishes to infinite order at $x = 0$ we can subtract the Taylor series of order $m-1$

$$T_2 g(x) = |x|^{-t}f(x) = |x|^{-t} \, [\, f(x) - \sum_{|\alpha| \leqslant m-1} f^{(\alpha)}(0)x^\alpha/\alpha!]$$

$$= |x|^{-t}\int [\, c_2|x-y|^{2-n} - \sum_{|\alpha| \leqslant m-1} a_\alpha(y)x^\alpha \,] \; g(y)|y|^{t-2}dy,$$

in which $\Sigma a_\alpha(y)x^\alpha$ is the Taylor polynomial for $c_2|x-y|^{2-n}$ about $x = 0$ with y as a parameter. It is easy to see that the *only* appropriate choice of Taylor polynomial is the one we have made if we want an inequality of the form (*') for *arbitrary* smooth functions g supported in $1 < |x| < 2$ as opposed to those that can be written in the form $|x|^{-t+2}\Delta f(x)$ (which satisfy additional moment conditions). We can now define the analytic continuation as (T_z depends implicitly on t) :

$$T_z g(x) = \int K_z(x,y) g(y) |y|^{-n} dy$$

with

$$K_z(x,y) = |x|^{-t} [c_z |x-y|^{z-n} - \sum_{|\alpha| \leqslant m-1} a_{\alpha,z}(y) x^\alpha] |y|^{t-z+n}$$

and with $\Sigma a_{\alpha,z}(y) x^\alpha$ the Taylor polynomial for $c_z|x-y|^{z-n}$ about $x = 0$ with y as a parameter.

Theorem 3 is proved by complex interpolation [15] between estimates for T_z on the vertical lines $\mathrm{Re} z = 0$ and $\mathrm{Re} z = n$. When $\mathrm{Re} z = 0$, we have singular integral kernels ($|x - y|^{i\gamma - n}$, $\gamma \in \mathbb{R}$) so that it is natural to expect an estimate $L^2 \to L^2$.

<u>LEMMA 1</u>. There exists a constant C depending only on δ and n such that for every $\gamma \in R$ and every $g \in C_0^\infty(\mathbb{R}^n \setminus \{0\})$ we have

$$\|T_{i\gamma} g\|_{L^2(\mathbb{R}^n, \frac{dx}{|x|^n})} \leqslant C e^{C|\gamma|} \|g\|_{L^2(\mathbb{R}^n, \frac{dx}{|x|^n})}.$$

When $\mathrm{Re} z = n$ we have bounded kernels ($|x - y|^{i\gamma}$, $\gamma \in \mathbb{R}$) but the pole in the coefficient c_z gives a logarithmic singularity at $\mathrm{Re} z = n$, the kernel for $\Delta^{-n/2}$ is a constant times $\log |x - y|$. As a result, we want to formulate a result close to an $L^1 \to L^\infty$ estimate, but allowing logarithmic singularities. As it turns out we only need to enlarge the space L^∞ in the radial direction. For a function f(x) on $\mathbb{R}^n$ we write polar coordinates $r = |x|$ and $\omega = x/|x|$ and the Mellin transform $\tilde{f}(n,\omega) = \int_0^\infty f(r\omega) r^{-i\eta - 1} dr$ in the r variable with ω fixed.

<u>LEMMA 2</u>. There exists an absolute constant C such that if $t \geqslant 1/2$ then for every $\gamma \in R$ and every $g \in C_0^\infty(R^n \setminus \{0\})$ we have

$$\sup_{\omega} \sup_{\eta \in \mathbb{R}} |\eta| \, |(T_{n+i\gamma} g)^\sim(\eta,\omega)| \leqslant C e^{C|\gamma|} \|g\|_{L^1(\mathbb{R}^n, \frac{dx}{|x|^n})} .$$

(The supremum in ω is over S^{n-1}, the unit sphere in $\mathbb{R}^n$.) ■

In order to explain the rather complicated-looking norm on the left-hand side of this last inequality, let us interpret all our space in terms of polar coordinates. Denote

$$Y_p = L^p(\mathbb{R}_+, dr/r) \quad \text{for } 1 \leq p < \infty$$

and

$$Y_\infty = \{h \in L^1_{loc}(\mathbb{R}_+, dr/r) : \sup_\eta |\eta||\tilde{h}(\eta)| < \infty\}.$$

Notice that

$$\int_{\mathbb{R}^n} |f(x)|^p |x|^{-n} dx = \int_{S^{n-1}} \int_0^\infty |f(r\omega)|^p r^{-1} dr\, d\omega.$$

Thus the space $L^p(S^{n-1}, d\omega; Y_p)$ of p-integrable vector-valued functions on the sphere with values in Y_p is isomorphic to $L^p(\mathbb{R}^n, |x|^{-n}dx)$. Now we see that Lemma 1 says that $T_{i\gamma}$ is bounded from $L^2(\mathbb{R}^n, |x|^{-n}dx)$ to $L^2(S^{n-1}, d\omega;\, Y_2)$ and Lemma 2 says that $T_{n+i\gamma}$ is bounded from $L^1(\mathbb{R}^n, |x|^{-n}dx)$ to $L^\infty(S^{n-1}, d\omega; Y_\infty)$. A slight variant of Stein's interpolation theorem for analytic families of operators then shows that for $0 < \theta < 1$

$T_{\theta n}$ is bounded from $L^{q(\theta)}(\mathbb{R}^n, |x|^{-n}dx)$ to

$$L^{p(\theta)}(S^{n-1}, d\omega; Y_{p(\theta)}) = L^{p(\theta)}(\mathbb{R}^n, |x|^{-n}dx),$$

where $p(\theta)$ and $q(\theta)$ are given by the formulas

$$1/q(\theta) = (1-\theta)/2 + \theta \text{ and } 1/p(\theta) = (1-\theta)/2 + \theta/\infty = (1-\theta)/2.$$

The point is that the space Y_∞ behaves like L^∞ under interpolation. Theorem 3 is just the special case $\theta = 2/n$.

Before passing to the proof of the lemmas, let us comment on the connection between the space Y_∞ and BMO. Recall that $\mathrm{BMO}(\mathbb{R}) = \{f \in L^1_{loc}(\mathbb{R}) :$ $\sup_I \frac{1}{|I|}\int_I |f(x) - f_I| dx < \infty$, where the supremum is over all intervals I and f_I denotes the average of f over I$\}$. The dual statement to Hardy's inequality

[17, vol 1 p. 286 (8.8)] is the fact that if

$$\sup_{\xi} |\xi| \, |\hat{f}(\xi)| < \infty,$$

then

$$f \in BMO(\mathbb{R}).$$

(Here $\hat{f}$ denotes the Fourier transform on $\mathbb{R}$.)
Therefore,

$$Y_\infty \subset \{h \in L^1_{loc}(\mathbb{R}_+, dr/r) \ : \ h(e^x) \in BMO(\mathbb{R})\} \ .$$

The space BMO also behaves like L^∞ under complex interpolation, but this theorem is much harder [4] . Fortunately, the function $f(x) = \log|x|, x \in \mathbb{R}$ belongs to the smaller space – $\hat{f}(\xi)$ is a constant times $|\xi|^{-1}$ – so that the theorem does not depend on such heavy machinery. In fact, the simplification due to Stein obviates even the space Y_∞, but we feel that it may come up in other contexts.

The proof of Lemma 1 depends on the following formulas .
Let $r > 0$, $\omega \in S^{n-1}$, $h \in C_0^\infty(\mathbb{R}_+ \setminus \{0\})$ and $P_k(\omega)$ a harmonic polynomial on $\mathbb{R}^n$, homogeneous of degree k (i.e. a spherical harmonic). Put $g(r\omega) = h(r)P_k(\omega)$ and $\zeta = t + i\eta$. The operator T_z (which depends implicity on t) can be written

$$T_z g(r\omega) = \int_0^\infty R_{k,t,z}(r/s)h(s)s^{-1} ds \ P_k(\omega),,$$

where

$$\tilde{R}_{k,t,z}(\eta) = \frac{2^{-z}\Gamma(\frac{1}{2}(k-\zeta))\Gamma(\frac{1}{2}(n+k-z+\zeta))}{\Gamma(\frac{1}{2}(k-\zeta+z))\Gamma(\frac{1}{2}(n+k+\zeta))}$$

The kernel $R_{k,t,s}$ acts on h as a convolution operator in the group $\mathbb{R}_+$ under multiplication (with Haar measure dr/r). The Mellin transform is the analogue of the Fourier transform in this setting, and the norm of T_z as an operator on $L^2(S^{n-1}, d\omega; Y_2)$ is just $\sup_{\eta \in \mathbb{R}, k \in \mathbb{N}} |\tilde{R}_{k,t,z}(\eta)|$. We can now read off the desired bound from Stirling's formula and the functional equation $\Gamma(\zeta)\Gamma(1-\zeta) = \pi/\sin \pi\zeta$. The poles of the gamma function show why integer values of t are not allowed. They can be used to construct counterexamples to Lemma 1 and Theorem 3 in the case of integer values of t.

To prove the formula we use the ordinary Fourier transform and inverse Fourier transform on R^n applied to the function $|x|^{t-2}h(|x|)P_k(\omega)$, multiply by $|\xi|^{-z/2}$ where ξ is the dual variable to x, and take the inverse Fourier transform. In this way the function $R_{k,t,z}$ is expressed as an integral involving two Bessel functions. A formula of Weber and Schaftheitlein then expresses $R_{k,t,z}$ in terms of hypergeometric functions. By what might seem like astounding luck, this particular combination of hypergeometric functions has a very simple Mellin transform. However, by hindsight one can see that there is considerable hope for a simple formula in light of the special case of even integer values of z, that is, integer powers of the Laplacian. In particular, one can see that it is natural that the function $R_{k,t,z}$ be a meromorphic function of the complex variable $\zeta = t + i\eta$.

There is one further element of the proof of this formula that deserves mention. The kernel $R_{k,t,z}$ is actually rather complicated because it involves the subtraction of the Taylor series of order $m-1$ involved in the definition of T_z. However the Mellin transform is ideally suited to take into account this subtracted piece. More precisely we have

REMARK. Suppose that $f \in C^\infty([0,\infty))$ and has compact support. The function $F(\lambda) = \int_0^\infty f(r) r^{-\lambda-1} dr$, initially defined for $\mathrm{Re}\lambda < 0$, has a unique meromorphic extension to the whole complex plane with (at most) simple poles at non-negative integers. The residue at the pole $\lambda = m$ is $-f^{(m)}(0)/m!$ and the extension, which we continue to denote by F has the formula

$$F(\lambda) = \int_0^\infty \Big(f(r) - \sum_{j=0}^{m-1} \frac{1}{j!} r^j f^{(j)}(0)\Big) r^{-\lambda-1} dr.$$

whenever $m-1 < \mathrm{Re}\lambda < m$. ■

This remark is easy to prove by integration by parts. It shows that one can prove the formula for $R_{k,t,z}(\eta)$ for certain $\zeta = t + i\eta$ and then use analytic continuation in ζ.

In the proof of Lemma 2 we still use polar coordinates and the Mellin transform, but spherical harmonic expansions are no longer helpful. It is easy to see by rotational symmetry that for $g(x) = g(r\omega)$

$$(T_z g)^{\sim}(\eta,\omega) = \int_{S^{n-1}} A_{z,t}(\eta,\omega.\omega')\tilde{g}(\eta,\omega')d\omega'$$

where $\omega.\omega'$ denotes the inner product in $\mathbb{R}^n$ of the unit vectors ω and ω'. It is not hard to see that

$$A_{z,t}(\eta,\alpha) = \text{const.} \int_0^{\infty}(1-2r\alpha+r^2)^{\frac{z-n}{2}} r^{-\zeta-1}dr$$

with

$$\zeta = t+i\eta, \ -1 \leqslant \alpha \leqslant 1.$$

(The integral is initially defined for $-1 < \mathrm{Re}(z-n)/2 < \mathrm{Re}\zeta < 0$ and continued analytically using the remark above.)
The lemma now follows immediately from the estimate

$$\sup_{\eta\in\mathbb{R},\, t\geqslant\frac{1}{2},\, -1\leqslant\alpha\leqslant 1} |\eta| \ A_{n+i\gamma,t}(\eta,\gamma) \leqslant Ce^{C|\gamma|} \qquad (**)$$

for some absolute constant C and all $\gamma \in \mathbb{R}$.
The final estimate (**) is nearly an old result of G.N. Watson. The function given here is a Gegenbauer function, and thus it can be expressed in terms of hypergeometric functions. Watson [16] gave asymptotic expansions for a large class of hypergeometric functions that imply in this case the bound (**) for α in a compact subinterval of $-1 < \alpha < 1$. However, if one repeats his (extremely tedious) steepest descent argument one finds (exactly in the case $\mathrm{Re}z = n$) that the bound (**) is uniform in the full interval $-1 \leqslant \alpha < 1$. This concludes the proof sketch.

E.M. Stein has recently simplified the proof considerably by replacing Lemma 2 with the following

LEMMA 3. If $n-1 < \mathrm{Re}\,z < n$ and $\gamma = \mathrm{Im}\,z$, then

$$\|T_z g\|_{L^p(\mathbb{R}^n, \frac{dx}{|x|^n})} \leqslant Ce^{C|\gamma|}\|g\|_{L^q(\mathbb{R}^n, \frac{dx}{|x|^n})},$$

where

$$g \in C_0^\infty(\mathbb{R}^n \setminus \{0\}),$$

C depends only on δ, n, p, q with $\frac{1}{q} - \frac{1}{p} = \frac{\mathrm{Re}\,z}{n}$, $1 < q < p < \infty$. ∎

He proves this result by a direct, fairly elementary estimate on the kernel. Interpolation then gives Theorem 3. Better still, Theorem 3 is also valid for values of p and q satisfying $1/q - 1/p = 2/n$ in a neighborhood of the values $q = 2n/(n+2)$ and $p = 2n/(n-2)$. It then follows from real interpolation of the single operator T_2 that T_2 is bounded from $L^q(\mathbb{R}^n, \frac{dx}{|x|^n})$ to the Lorentz space $L^{p,q}(\mathbb{R}^n, \frac{dx}{|x|^n})$ with $q = 2n/(n+2)$ and $p = 2n/(n-2)$. (see [15] for a definition of Lorentz spaces). Finally, this permits us to extend Theorem 2 to the case in which a(x) belongs to weak $L^{n/2}$ and has small weak $L^{n/2}$ norm on balls of small radius.

References

[1] S. Agmon, Lower bounds for Schrödinger equations. J. d'Analyse 23 (1970), 1-25.

[2] W.O. Amrein, A.M. Berthier, and V. Georgescu, L^p inequalities for the Laplacian and unique continuation, Ann. Inst. Fourier (Grenoble) 31 (1981), 153-168.

[3] T. Carleman, Sur un problème d'unicité pour les systèmes d'équations aux dérivées partielles à deux variables indépendentes, Ark. Mat 26B (1939), 1-9.

[4] C. Fefferman and E.M. Stein, H^p spaces of several variables, Acta Math. 129 (1972), 137-193.

[5] R. Froese, J. Herbst, M. and T. Hoffman-Ostenhof, On the absence of positive eigenvalues for one-body Schrödinger operators. J. d'Analyse, 41 (1982) 272-284.

[6] V. Georgescu, On the unique continuation property for Schrödinger Hamiltonians, Helv. Acta 52 (1979), 655-670.

[7] L. Hormander, Linear partial differential operators, Springer-Verlag Berlin and New York (1964).

[8] L. Hormander, Uniqueness theorems for second order elliptic differential equations, Comm. in Partial Differential Equations, 8 (1) (1983), 21-64.

[9] D. Jerison and C.E. Kenig, Unique continuation and absence of positive eigenvalues for Schrödinger operators. To appear Annals of Math.

[10] J. Saut and B. Scheurer, Un théorème de prolongement unique pour des opérateurs elliptiques dont les coefficients ne sont pas localement bornés. C.R. Acad. Sci. Paris Ser A 290 (1980), 598-599.

[11] E. Sawyer, Unique continuation for Schrödinger operators in dimension three or less, preprint.

[12] M. Schechter and B. Simon, Unique continuation for Schrödinger operators with unbounded potential. J. Math. Anal. Appl. 77 (1980), 482-292.

[13] B. Simon, Schrödinger semigroups, Bull AMS 7(3) (1983), 447-526.

[14] E. M. Stein, Singular integrals and differentiability properties of functions, Princeton U. Press, Princeton, N.J. (1970).

[15] E.M. Stein and G. Weiss, Introduction to Fourier Analysis on Euclidean space, Princeton U. Press, Princeton, N.J. (1971).

[16] G.N. Watson, Asymptotics of hypergeometric functions, Trans. Cambridge Philos. Soc. 22 (1918), 277-308.

[17] A. Zygmund, Trigonometric Series, 2nd Ed., Cambridge U. Press, London and New York (1968).

The author was supported in part by NSF Grant MCS - 8202127.

David Jerison
Department of Mathematics
Massachusetts Institute of Technology
Cambridge, MA 02139
U.S.A.

C E KENIG

Boundary value problems on Lipschitz domains

The purpose of this note is to describe some of the highlights of the recent progress in the study of elliptic boundary value problems on Lipschitz domains. Expository accounts of the results that I am going to expound upon can be found in [11], [20], [21], and [22]. We will refer to the original articles in the text.

In the late 50's and early to mid 60's, the study of elliptic problems with L^p data on smoothly bounded domains became well understood. This was due to the work of A.P. Calderon and A. Zygmund on multidimensional singular integrals ([5], [6]), the work of Agmon, Douglis and Nirenberg ([1], [2]) on *a priori* estimates, the development of the theory of pseudo differential operators by Kohn and Nirenberg [23], and the applications by A.P. Calderon [3] and R. Seeley [27] to the reduction of elliptic boundary value problems to pseudo differential equations on the boundary. This program constituted a far reaching extension of the classical method of layer potentials introduced at the beginning of the century by C. Neumann, and later studied by Fredholm, Giraud, Mikhlin and others. In this manner, optimal estimates were obtained for general elliptic boundary value problems on smooth domains, with L^p data.

However, if one goes back to the physical and geometrical origin of the elliptic boundary value problems, it becomes evident that the domains that appear naturally are not smooth, but have faces, edges and vertices. A good class of domains that allows for this kind of roughness is the class of Lipschitz domains. These are bounded domains in $\mathbb{R}^n$ which are given locally as the domain above the graph of a Lipschitz function $\varphi : \mathbb{R}^{n-1} \to \mathbb{R}$. This class of domains has the very important property that it remains invariant (together with the natural bounds associated to it) under dilations. This is not true for the class of smooth domains, and this is one of the reasons why Lipschitz domains are very important from the mathematical point of view.

In order to discuss the work that has been done on the boundary value problems on Lipschitz domains, we will start by describing the results on the Dirichlet and Neumann problem for Laplace's equation.

In a series of papers between 1977 and 1978, B. Dahlberg ([8], [9], [10]) was able to solve, with optimal estimates, the Dirichlet problem for Laplace's equation in a bounded Lipschitz domain D, with boundary data $f \in L^p(\partial D)$. He was able to show that there exists a unique function u with

$$\begin{aligned} &\Delta u = 0 \text{ in } D \\ &u\big|_{\partial D} = f \end{aligned} \qquad (1)$$

and the non-tangential maximal function $u^*(Q) = \sup_{X \in \Gamma_\alpha(Q)} |u(X)|$,

$\Gamma_\alpha(Q) = \{X \in D : \text{dist}(X, \partial D) < (1+\alpha)|X-Q|\}$, is in $L^p(\partial D)$. This holds in the range $2-\varepsilon < p \leqslant \infty$, $\varepsilon = \varepsilon(D) > 0$. The range of p's is sharp as simple examples show, and the equality $u_{|\partial D} = f$ holds in the sense that

$$\lim_{\substack{X \to Q \\ X \in \Gamma_\alpha(Q)}} u(X) = f(Q) \text{ for a.e. } Q.$$

We will abreviate this statement by saying that $u_{|\partial D} = f$ non-tangentially a.e. He also obtained Sobolev and Besov space estimates. For example he showed that if $f \in L^2(\partial D)$, then $u \in H^{1/2}(D)$. Finally he showed that if D is a C^1 domain, the results hold for $1 < p < \infty$.

Dahlberg's approach consisted in analyzing in great detail the harmonic measure for D. His methods relied on clever applications of the maximum principle, the Harnack principle, and the theory of weighted norm inequalities of Muckenhoupt. The disadvantage of Dahlberg's method is that it relies on positivity, and this precluded its applicability to the Neumann problem, or to the Dirichlet problem for systems of second order equations, or to higher order problems.

In a related development, in 1978, Fabes, Jodeit and Riviere ([15]) were able to extend to C^1 domains the classical method of layer potentials, to solve the Dirichlet and Neumann problem for Laplace's equation.

Let us recall this classical method to study the Dirichlet problem (1). Following C. Neumann, one writes

$$u(X) = \frac{1}{\omega_n} \int_{\partial D} \frac{\langle X-Q, N_Q \rangle}{|X-Q|^n} g(Q)\, d\sigma(Q),$$

where N_Q denotes the unit normal at $Q \in \partial D$, and $\langle , \rangle$ the inner product in $\mathbb{R}^n$. It is easy to see that u is harmonic in D, and that, if D is smooth,

$$u|_{\partial D} = \frac{1}{2} g + \frac{1}{\omega_n} \int_{\partial D} \frac{\langle P-Q, N_Q \rangle}{|P-Q|^n} g(Q)\, d\sigma(Q) = (\frac{1}{2} I + K)(g).$$

If ∂D is of class $C^{1,\alpha}$, one has the estimate $\left| \frac{\langle P-Q, N_Q \rangle}{|P-Q|^n} \right| \leqslant \frac{C}{|P-Q|^{n-1+\alpha}}$,

which shows that K is L^p bounded and compact, $1 < p < \infty$. The Fredholm theory then allows one to solve the equation $(\frac{1}{2} I + K)(g) = f$, and (1) is thus solved. This is the method extended to the general theory of elliptic boundary value problems by the theory of singular integrals and pseudo differential operators. What happens when ∂D is merely C^1 or Lipschitz ? In this case the L^p boundedness of K is even in question. In 1977, A.P. Calderon ([4]) was able to show that K is L^p bounded for C^1 domains. In [15], Fabes, Jodeit and Riviere observed that Calderon's proof actually could be used to show the compactness in $L^p(\partial D)$, $1 < p < \infty$ of K. They were then able to use the Fredholm theory to solve the Dirichlet problem (1) in $L^p(\partial D)$, $1 < p < \infty$, for C^1 domains, and also the Neumann problem

$$\Delta u = 0 \text{ in } D$$

$$\left.\frac{\partial u}{\partial N}\right|_{\partial D} = f \text{ on } \partial D \tag{2}$$

where $\frac{\partial u}{\partial N}$ is the normal derivative, and the second equality holds non-tangentially a.e. on ∂D, and $(\nabla u)^* \in L^p(\partial D)$, $1 < p < \infty$, for D a C^1 domain.

Moreover, they found representation formulas for the solutions of (1) and (2), in terms of classical layer potentials.

The difficulty in showing the L^p boundedness of K for C^1 or Lipschitz domains resides in the case p = 2, where K is a singular integral operator which is not sufficiently close to one of convolution type to allow for the use of the Fourier transform. Whether K was L^p bounded for Lipschitz domains remained open at this point.

In 1979, D. Jerison and C. Kenig ([17], [18]) gave a new, simpler proof of Dahlberg's results [8], using an integral identity, which is a consequence of the divergence theorem. This identity goes back to Rellich ([26], 1940) and was used by various authors in different contexts (see for example [25], [24]). This identity easily implies that if D is a Lipschitz domain, and $\Delta u = 0$ in D, then

$$\int_{\partial D} |\nabla_t u|^2 d\sigma \approx \int_{\partial D} \left|\frac{\partial u}{\partial N}\right|^2 d\sigma, \tag{3}$$

where $\nabla_t u$ denotes the tangential part of ∇u. Shortly after, (1980) D. Jerison and C. Kenig ([19]) were able to use (3) to treat the Neumann problem (2) for Lipschitz domains with L^2 data, reducing it to Dahlberg's L^2 results for problem (1). They also showed that the solution belongs to the Sobolev space $H^{3/2}(D)$. The approach of Jerison and Kenig for (1) and the L^2 case of (2) still relied on positivity, which precluded the extension to systems of equations. Also no good representation formula for the solutions existed, and the L^p case of (2) for Lipschitz domains remained open.

In 1981, R. Coifman, A. McIntosh and Y. Meyer ([7]) established the L^p boundedness of the Cauchy integral on any Lipschitz curve, and as a consequence, the L^p boundedness, $1 < p < \infty$ of the operator K mentioned above for any Lipschitz domain in $\mathbb{R}^n$. This opened the door to the applicability of the method of layer potentials to Lipschitz domains. This method is very flexible, does not rely on positivity, and does not in principle differentiate between a single equation or systems of equations. The difficulty that arises in the case of Lipschitz domains resides in the fact that K need not be compact, as simple examples show, and hence the Fredholm theory does not apply to invert $(\frac{1}{2} I + K)$. In 1982, in his doctoral dissertation, G. Verchota ([28]) showed that, in the case of Lipschitz domains, the estimate (3) which had been previously obtained by Jerison and Kenig ([19]), is the appropriate substitute for compactness, in the $L^2(\partial D)$ case. He was thus able to recover the L^2 results of Dahlberg ([9]) on (1), and the L^2 results of Jerison and Kenig ([19]) on (2), but also obtaining representation formulas in terms of layer potentials.

What remained open at this point was recovering Dahlberg's L^p results for (1) using layer potentials, and obtaining the sharp L^p results for (2) on Lipschitz domains. Both of these questions have been recently resolved by Dahlberg and Kenig ([12]). It was shown in [12] that given a Lipschitz domain $D \subset \mathbb{R}^n$, there exists $\varepsilon = \varepsilon(D) > 0$ such that one can solve the Neumann problem (2) with L^p data, $1 < p < 2+\varepsilon$. Easy examples show that the range of p's is optimal. It was also shown that the solution can be obtained by the method of layer potentials, and that the same is true for the solution of (1) in the optimal range of p's. The key idea in this work is to use the so called atoms that arise in the theory of Hardy spaces, together with the De Giorgi-Nash-Moser theory of regularity of solutions of equations with measurable coefficients.

Also, very recently it has been possible to extend the L^2-results of Verchota ([28]) to second order systems of elliptic equations. For example, Dahlberg, Kenig and Verchota ([13]) have been able to solve by the method of layer potentials, and with optimal estimates on arbitrary Lipschitz domains in $\mathbb{R}^n$, the following problems in the mathematical theory of elasticity :

$$\begin{aligned} &\mu\Delta\vec{u} + (\lambda+\mu)\nabla \operatorname{div} \vec{u} = 0 \text{ in } D \\ &\vec{u}\big|_{\partial D} = f \in L^2(\partial D) \end{aligned} \tag{4}$$

$$\begin{aligned} &\mu\Delta\vec{u} + (\lambda+\mu)\nabla \operatorname{div} \vec{u} = 0 \text{ in } D \\ &\lambda(\operatorname{div} u)N + \mu\{\nabla u + \nabla u^t\}N = f \in L^2(\partial D). \end{aligned} \tag{5}$$

Here $\mu > 0$, $\lambda \geqslant 0$, and $\vec{u} : D \to \mathbb{R}^n$.

Also, Fabes, Kenig and Verchota [16] have been able to solve, by the method of layer potentials on arbitrary Lipschitz domains, the Stokes problem of hydrostatics,

$$\begin{aligned} &\Delta \vec{u} = \nabla p \text{ in } D \\ &\operatorname{div} \vec{u} = 0 \text{ in } D \\ &\vec{u}\big|_{\partial D} = \vec{f} \in L^2(\partial D). \end{aligned} \tag{6}$$

To conclude this survey, we would like to mention the following result of Dahlberg, Kenig and Verchota ([14]), on a fourth order problem on a Lipschitz domain :

$$\begin{aligned} &\Delta^2 u = 0 \text{ in } D \\ &u\big|_{\partial D} = f \in H^1(\partial D) \\ &\frac{\partial u}{\partial N}\Big|_{\partial D} = g \in L^2(\partial D). \end{aligned} \tag{7}$$

The authors have obtained a unique solution in the class $H^{3/2}(D)$, which is the optimal estimate.

Many problems in this area remain open. I would like to mention the possibility of a general theory to treat all elliptic boundary value problems on Lipschitz domains, with optimal estimates, in a systematic way. Another problem of great interest is the treatment of the parabolic versions of the problems discussed in this note. Nothing is known in this direction. It would be very interesting to be able to study also non-linear versions of these problems, such as the Navier-Stokes equations, in Lipschitz cylinders.

References

[1] S. Agmon, A. Douglis and L. Nirenberg, Estimates near the boundary for solutions of elliptic partial differential equations satisfying general boundary conditions, I, Comm. Pure Appl. Math. 12, 623-727 (1959).

[2] S. Agmon, A. Douglis and L. Nirenberg, Estimates near the boundary for solutions of elliptic partial differential equations satisfying general boundary conditions, II, Comm. Pure Appl. Math. 17, 35-92 (1964).

[3] A.P. Calderon, Boundary value problems for elliptic equations, Outlines of the joint Soviet-American symposium on partial differential equations, 303-304, Novosibirsk (1963).

[4] A.P. Calderon, Cauchy integrals on Lipschitz curves and related operators, Proc. Nat. Acad. of Sci. U.S.A., 74, 1324-1327 (1977).

[5] A.P. Calderon and A. Zygmund, On the existence of certain singular integrals, Acta Math. 88, 85-139 (1952).

[6] A.P. Calderon and A. Zygmund, Singular integral operators and differential equations, Amer. J. of Math. 79, 289-309 (1957).

[7] R.R. Coifman, A. McIntosh and Y. Meyer, L'intégrale de Cauchy définit un opérateur borné sur L^2 pour les courbes lipschitziennes, Annals of Math. 116 (1982), 361-387.

[8] B.E.J. Dahlberg, On estimates of harmonic measure, Arch. Rat. Mech. and Anal. 65, 272-288 (1977).

[9] B.E.J. Dahlberg, On the Poisson integral for Lipschitz and C^1 domains, Studia Math. 66, 13-24 (1979).

[10] B.E.J. Dahlberg, Weighted norm inequalities for the Lusin area integral and the non-tangential maximal function for functions harmonic in a Lipschitz domain, Studia Math. 67, 297-314 (1980).

[11] B.E.J. Dhlberg, Harmonic functions in Lipschitz domains, Proc. Symp. in Pure Math., Vol XXXV, Part 1, 313-322 (1979).

[12] B.E.J. Dahlberg and C.E. Kenig, Hardy spaces and the L^p Neumann problem for Laplace's equation in a Lipschitz domain, preprint.

[13] B.E.J. Dahlberg, C.E. Kenig and G.C. Verchota, Boundary value problems for the systems of elastostatics in a Lipschitz domain, in preparation.

[14] B.E.J. Dahlberg, C.E. Kenig and G.C. Verchota, The Dirichlet problem for the biharmonic equation in a Lipschitz domain, to appear, Annales de l'Institut Fourier, Grenoble.

[15] E. Fabes, M. Jodeit Jr. and N.M. Riviere, Potential techniques for boundary value problems on C^1 domains, Acta Math. 141 (1978), 165-186.

[16] E. Fabes, C. Kenig and G. Verchota, The Stokes system in a Lipschitz domain, in preparation.

[17] D.S. Jerison and C.E. Kenig, An identity with applications to harmonic measure, Bull. AMS Vol 2 (1980), 447-451.

[18] D.S. Jerison and C.E. Kenig, The Dirichlet problem in non-smooth domains, Annals of Math. 113 (1981), 367-382.

[19] D.S. Jerison and C.E. King, The Neumann problem on Lipschitz domains, Bull. AMS Vol 4 (1981), 203-207.

[20] D.S. Jerison and C.E. Kenig, Boundary value problems on Lipschitz domains, MAA Studies in Mathematics, Vol 23, Studies in Partial Differential Equations, W. Litman, editor (1982), 1-68.

[21] C.E. Kenig, Recent progress on boundary value problems on Lipschitz domains, to appear, Proc. of Symp. in Pure Math., Proc. of the Notre Dame Conference on Pseudodifferential Operators.

[22] C.E. Kenig, Elliptic boundary value problems on Lipschitz domains, to appear in Beijing Lectures in harmonic analysis, Annals of Math. Studies, Princeton University Press.

[23] J.J. Kohn and L. Nirenberg, On the algebra of pseudo-differential operators, Comm. Pure Appl. Math. 18, 269-305 (1965).

[24] J. Nečas, Les méthodes directes en théorie des équations elliptiques, Academia, Prague (1967).

[25] L. Payne and H. Weinberg, New bounds in harmonic and biharmonic problems, J. Math. Phys. 33 (1954), 291-307.

[26] F. Rellich, Darstellung der Eigenwerte von $\Delta u + \lambda u$ durch ein Randintegral, Math. Z. 46 (1940), 635-646.

[27] R. Seely, Singular integrals and boundary problems, Amer. J. Math. 88, 781-809 (1966).

[28] G.C. Verchota, Layer potentials and boundary value problems for Laplace's equation in Lipschitz domains, Thesis, University of Minnesota (1982), also to appear in J. of Functional Analysis.

Carlos E. Kenig
School of Mathematics
University of Minnesota
Minneapolis, MN 55455
U.S.A.

O A OLEINIK, A S SHAMAEV & G A YOSIFIAN

Asymptotic expansions of solutions of the Dirichlet problem for elliptic equations in perforated domains

In this paper we consider asymptotic expansions for solutions of the Dirichlet boundary value problem for elliptic equations and systems in domains with ε-periodically situated cavities or cracks. Such domains are called perforated or domains with a finely granulated boundary. There is a vast literature on the asymptotic analysis of boundary value problems in perforated domains (see, for example [1] -[9]).

First results on the asymptotic expansion of solutions of the Dirichlet problem for the equation $\Delta U_\varepsilon = f(x)$ in a perforated domain Ω_ε were obtained by J.L. Lions in [2] together with the estimates for the remainder term, provided that f(x) is a function with a compact support in Ω. The Neumann problem is considered in [3] - [9].

In the present paper we construct asymptotic expansions and prove estimates of the remainder term for solutions of the Dirichlet problems related to the equations $\Delta U_\varepsilon = f$, $\Delta^2 U_\varepsilon = f$ and to the system of linear elasticity, without the assumption that f(x) has a compact support in Ω. All results obtained in this paper can also be proved in the same way for any second order elliptic equations of divergent form with bounded measurable coefficients, as well as for strongly elliptic equations of higher order.

To prove the validity of the asymptotic expansion of solutions for an arbitrary $f(x) \in C^\infty$ one has to construct some boundary layer functions, which decay exponentially with the increase of the distance from the boundary.

In the case of the Neumanntype problem for the system of linear elasticity in domains of some special form the boundary layer functions are constructed in [10], and for second order elliptic equations in [11].

Some results obtained in the present paper can be extended to the case of the mixed conditions of Dirichlet and Neumann type on the boundary of the cavities, as well as to the case of some non-periodic structures (see section 4).

1. Second order elliptic equations

For simplicity we consider first the Dirichlet problem for the Laplace equation. Let G^0 be a non-empty closed set belonging to the unit cube $Q = \{\xi \in \mathbb{R}^n, 0 \leqslant \xi_j \leqslant 1, j = 1,\ldots,n\}$. Let $G_1 = \bigcup_{z \in \mathbb{Z}^n} (G^0 + z)$, where $\mathbb{Z}^n$ is the set of all vectors $z = (z_1,\ldots,z_n)$ with integer components. We use the notation $G_\varepsilon = \{x : \varepsilon^{-1} x = \xi \in G_1\}$, $\Omega_\varepsilon = \Omega \setminus G_\varepsilon$, where Ω is a domain in $\mathbb{R}^n$. It is assumed that the sets $\mathbb{R}^n \setminus G_1$ and Ω_ε are connected.

In Ω_ε we shall obtain the asymptotic expansion in powers of the small parameter ε of the solution of the following boundary value problem

$$\begin{aligned} &\Delta u_\varepsilon(x) = f(x) \quad \text{in } \Omega_\varepsilon, \\ &u = \Phi \quad \text{on } \partial\Omega_\varepsilon . \end{aligned} \tag{1}$$

We denote by $H^1(\omega)$ the completion of the space $C^1(\bar{\omega})$ with respect to the norm

$$\|v\|_{1,\omega} = \left(\int_\omega (|\nabla v|^2 + |v|^2) dx \right)^{1/2},$$

where $\nabla v = (\partial v/\partial x_1,\ldots,\partial v/\partial x_n)$. By $H_0^1(\omega)$ we denote the closure in $H^1(\omega)$ of the subspace of $C^1(\bar{\omega})$ consisting of functions v vanishing in a neighbourhood of $\partial\omega$.

A weak solution of the problem

$$\begin{aligned} &\Delta u = f_0 + \frac{\partial f_k}{\partial x_k} \quad \text{in } \Omega_\varepsilon, \\ &u = \Phi \text{ on } \partial\Omega_\varepsilon, \ \Phi \in H^1(\Omega_\varepsilon), f_0, f_k \in L_2(\Omega_\varepsilon), \ k = 1,\ldots,n, \end{aligned} \tag{2}$$

is defined as a function $u(x)$ from $H^1(\Omega_\varepsilon)$, such that $W \equiv u - \Phi \in H_0^1(\Omega_\varepsilon)$ and the integral identity

$$-\int_{\Omega_\varepsilon} \frac{\partial W}{\partial x_k} \frac{\partial v}{\partial x_k} dx = \int_{\Omega_\varepsilon} \left(f_0 v - f_k \frac{\partial v}{\partial x_k}\right) dx + \int_{\Omega_\varepsilon} \frac{\partial \Phi}{\partial x_k} \frac{\partial v}{\partial x_k} dx \tag{3}$$

holds for any $v \in H_0^1(\Omega_\varepsilon)$. Here and in what follows the summation over repeated indices from 1 to n is assumed.

Denote by $\hat{H}^1_0(Q\setminus G^0)$ the completion in the norm $H^1(Q\setminus G^0)$ of the space of functions $v(\xi)$ continuously differentiable in $\mathbb{R}^n \setminus G_1$ vanishing in a neighbourhood of ∂G_1 and periodic in $\xi_1,\ldots,\xi_n$ with period 1 (1-periodic in ξ).

We say that $g(\xi)$ is a 1-periodic in ξ weak solution of the problem

$$\begin{aligned} &\Delta_\xi g(\xi) = F_0(\xi) + \frac{\partial F_j(\xi)}{\partial \xi_j} \quad \text{in } \mathbb{R}^n\setminus G_1, \\ &g = 0 \quad \text{on } \partial G_1, \end{aligned} \tag{4}$$

where $F_0, F_j \in L_2(Q\setminus G^0)$, $j = 1,\ldots,n$, and are 1-periodic in ξ, if $g\in \hat{H}^1_0(Q\setminus G^0)$ and the integral identity

$$-\int_{Q\setminus G^0} \frac{\partial g}{\partial \xi_j}\frac{\partial v}{\partial \xi_j}\,d\xi = \int_{Q\setminus G^0}(F_0 v - F_j\frac{\partial v}{\partial \xi_j})\,d\xi$$

holds for any $v\in \hat{H}^1_0(Q\setminus G^0)$.

The set G^0 is assumed throughout the paper to be such that a kind of Friedrichs inequality, namely

$$\int_{Q\setminus G^0} |v|^2 d\xi \leqslant M \int_{Q\setminus G^0} |\nabla_\xi v|^2 d\xi \tag{5}$$

holds for any $v(\xi)$ in $C^1(\bar{Q})$ vanishing in a neighbourhood of G^0. If the projection of G^0 on one of the coordinate hyper-planes contains a (n-1)-dimensional ball, then inequality (5) is valid, (see [12], [13]).

We first prove some auxiliary results.

LEMMA 1. A weak solution $u(x)$ of problem (2) exists and satisfies the following inequalities

$$\int_{\Omega_\varepsilon} |\nabla u|^2 dx \leqslant c_1(\varepsilon^2 \int_{\Omega_\varepsilon} f_0^2\,dx + \int_{\Omega_\varepsilon} f_i f_i dx + \int_{\Omega_\varepsilon} |\nabla\Phi|^2\,dx), \tag{6}$$

$$\int_{\Omega_\varepsilon} |u|^2\,dx \leqslant c_2(\varepsilon^4 \int_{\Omega_\varepsilon} f^2\,dx + \varepsilon^2 \int_{\Omega_\varepsilon} f_i f_i\,dx + \varepsilon^2 \int_{\Omega_\varepsilon}|\nabla\Phi|^2 dx + \int_{\Omega_\varepsilon} |\Phi|^2\,dx), \tag{7}$$

where constant c_1 and c_2 do not depend on ε, Φ, f_0, f_i, $i = 1,\dots,n$.

Proof. The existence of $u(x)$ follows from the Lax-Milgram lemma and the Friedrichs inequality for Ω_ε. By the definition of $u(x)$ the function $W \equiv u-\Phi$ belongs to $H_0^1(\Omega_\varepsilon)$. Extending W as zero to $\mathbb{R}^n \setminus \Omega_\varepsilon$ and applying (5) to $W(x)$ for $x = \varepsilon\xi$ we obtain the estimate

$$\int_{\Omega_\varepsilon} |W|^2 \, dx \leqslant M\varepsilon^2 \int_{\Omega_\varepsilon} |\nabla_x W|^2 \, dx. \tag{8}$$

Taking $v = W$ in the integral identity (3) we get

$$\int_{\Omega_\varepsilon} |\nabla W|^2 \, dx \leqslant \left| \int_{\Omega_\varepsilon} f_0 W dx \right| + \left| \int_{\Omega_\varepsilon} f_i \frac{\partial W}{\partial x_i} \, dx \right| + \left| \int_{\Omega_\varepsilon} \frac{\partial \Phi}{\partial x_i} \frac{\partial W}{\partial x_i} \, dx \right| .$$

It follows from (8) and Hölder's inequality that

$$\int_{\Omega_\varepsilon} |\nabla W|^2 \, dx \leqslant K(\varepsilon^2 \int_{\Omega_\varepsilon} f_0^2 \, dx + \int_{\Omega_\varepsilon} f_i f_i \, dx + \int_{\Omega_\varepsilon} |\nabla \Phi|^2 \, dx),$$

$$\int_{\Omega_\varepsilon} |W|^2 \, dx \leqslant KM(\varepsilon^4 \int_{\Omega_\varepsilon} f_0^2 \, dx + \varepsilon^2 \int_{\Omega_\varepsilon} f_i f_i dx + \varepsilon^2 \int_{\Omega_\varepsilon} |\nabla \Phi|^2 dx),$$

where constants K, M do not depend on ε. Therefore, since $W \equiv u-\Phi$, we have estimates (6), (7). The lemma is proved. □

First we obtain the formal asymptotic expansion in powers of ε of the solution $u_\varepsilon(x)$ of problem (1) with $\Phi \equiv 0$. Consider a formal series

$$\tilde{u}_\varepsilon(x) = \sum_{\ell=0}^{\infty} \varepsilon^{\ell+2} \sum_{\langle\alpha\rangle = \ell} N_\alpha^0(\xi) D^\alpha v(x), \quad \xi = \varepsilon^{-1} x, \tag{9}$$

where $N_\alpha^0(\xi)$, $v(x)$ are the functions to be determined,

$$\alpha = (\alpha_1, \dots, \alpha_\ell), \quad \alpha_j = 1, \dots, n, \quad D^\alpha = \frac{\partial^\ell}{\partial x_{\alpha_1} \dots \partial x_{\alpha_\ell}}, \quad \langle\alpha\rangle = \ell.$$

Substituting $\tilde{u}_\varepsilon(x)$ for $u_\varepsilon(x)$ in (1) we obtain a formal equality

$$\sum_{\ell=0}^{\infty} \varepsilon^{\ell} \sum_{\langle\alpha\rangle=\ell} H_{\alpha}(\xi) D^{\alpha} v(x) \cong f(x), \tag{10}$$

where

$$H_0(\xi) = \Delta_{\xi} N_0^0(\xi), \quad H_1(\xi) = \Delta_{\xi} N_{\alpha_1}^0(\xi) + 2 \frac{\partial N_0^0(\xi)}{\partial \xi_{\alpha_1}},$$

$$H_{\alpha_1 \ldots \alpha_s}(\xi) = \Delta_{\xi} N_{\alpha_1 \ldots \alpha}(\xi) + 2 \frac{\partial}{\partial \xi_{\alpha_1}} N^0_{\alpha_2 \ldots \alpha_s}(\xi) + \delta_{\alpha_1 \alpha_2} N^0_{\alpha_3 \ldots \alpha_s}, \quad s \geqslant 2,$$

$\delta_{\alpha_1\alpha_2} = 0$ for $\alpha_1 \neq \alpha_2$, $\delta_{\alpha_1\alpha_2} = 1$ for $\alpha_1 = \alpha_2$. Equating the coefficients by the same powers of ε in the right and left sides of (10) and setting $v(x) = f(x)$, we get for $N_{\alpha}^0(\xi)$ the following sequence of boundary value problems

$$\Delta_{\xi} N_0^0(\xi) = 1 \quad \text{in } \varepsilon^{-1}\Omega_{\varepsilon}, \quad N_0^0(\xi) = 0 \text{ on } \partial(\varepsilon^{-1}\Omega_{\varepsilon}), \tag{11}$$

$$\Delta_{\xi} N_{\alpha_1}^0(\xi) + 2 \frac{\partial N_0^0(\xi)}{\partial \xi_{\alpha_1}} = 0, \quad \text{in } \varepsilon^{-1}\Omega_{\varepsilon}, \quad N_{\alpha_1}^0(\xi) = 0 \quad \text{on } \partial(\varepsilon^{-1}\Omega_{\varepsilon}), \tag{12}$$

$$\Delta_{\xi} N^0_{\alpha_1 \ldots \alpha_s}(\xi) + 2 \frac{\partial}{\partial \xi_{\alpha_1}} N^0_{\alpha_2 \ldots \alpha_s}(\xi) + \delta_{\alpha_1 \alpha_1} N^0_{\alpha_3 \ldots \alpha_s} = 0 \text{ in } \varepsilon^{-1}\Omega_{\varepsilon},$$

$$N^0_{\alpha_1 \ldots \alpha_s}(\xi) = 0 \quad \text{on } \partial(\varepsilon^{-1}\Omega_{\varepsilon}), \quad s \geqslant 2. \tag{13}$$

It is easy to prove by induction that weak solutions $N_{\alpha}^0(\xi)$ exist, $N_{\alpha}^0(\frac{x}{\varepsilon}) \in H_0^1(\Omega_{\varepsilon})$.

We show now that $N_{\alpha}^0(\xi)$ can be written in the form $N_{\alpha}^0(\xi) = N_{\alpha}^1(\xi) + N_{\alpha}^2(\xi)$ where $N_{\alpha}^1(\xi)$ are 1-periodic in ξ functions belonging to $\hat{H}_0^1(Q \setminus G^0)$ and $N_{\alpha}^2(\varepsilon^{-1}x)$ are functions of boundary layer type. We set

$$T_0^1 = 1, \quad T_0^2 = 0, \quad T_{\alpha_1}^j(\xi) = -2 \frac{\partial}{\partial \xi_{\alpha_1}} N_0^j(\xi), \quad j = 1,2;$$

$$T^j_{\alpha_1\ldots\alpha_m}(\xi) = -2\frac{\partial}{\partial\xi_{\alpha_1}} N^j_{\alpha_2\ldots\alpha_m}(\xi) - \delta_{\alpha_1\alpha_2} N^j_{\alpha_3\ldots\alpha_m}(\xi),\ m>1.$$

Define $N^1_\alpha(\xi)$ as 1-periodic in ξ, weak solutions of the boundary value problems

$$\Delta_\xi N^1_\alpha(\xi) = T^1_\alpha(\xi) \quad \text{in } \mathbb{R}^n\setminus G_1,\ N^1_\alpha(\xi) = 0 \text{ on } \partial G_1, \tag{14}$$

and $N^2_\alpha(\xi)$ as weak solutions of the Dirichlet problems

$$\Delta_\xi N^2_\alpha(\xi) = T^2_\alpha(\xi) \quad \text{in } \varepsilon^{-1}\Omega_\varepsilon,\ N^2_\alpha = -N^1_\alpha \quad \text{on } \partial(\varepsilon^{-1}\Omega_\varepsilon). \tag{15}$$

LEMMA 2. Functions $N^j_\alpha(\varepsilon^{-1}x)$ satisfy the inequalities

$$\varepsilon^2\int_{\Omega_\varepsilon}|\nabla_x N^j_\alpha|^2dx + \int_{\Omega_\varepsilon}|N^j_\alpha|^2dx \leqslant M_\alpha,\ j = 0,1,2, \tag{16}$$

where constants M_α do not depend on ε.

Proof. It is easy to see that $N^0_\alpha(\varepsilon^{-1}x)$ are weak solutions of the following boundary value problems

$$\Delta_x N^0_0 = \varepsilon^{-2} \quad \text{in } \Omega_\varepsilon,\ N^0_0 = 0 \quad \text{on } \partial\Omega_\varepsilon,$$

$$\Delta_x N^0_{\alpha_1} = -2\varepsilon^{-1}\frac{\partial N^0_0}{\partial x_{\alpha_1}} \quad \text{in } \Omega_\varepsilon;\ N^0_{\alpha_1} = 0 \text{ on } \partial\Omega_\varepsilon,$$

$$\Delta_x N^0_{\alpha_1\ldots\alpha_s} = -2\varepsilon^{-1}\frac{\partial}{\partial x_{\alpha_1}} N^0_{\alpha_2\ldots\alpha_s} - \varepsilon^2\delta_{\alpha_1\alpha_2} N^0_{\alpha_3\ldots\alpha_s} \quad \text{in } \Omega_\varepsilon;$$

$$N^0_{\alpha_1\ldots\alpha_s} = 0 \text{ on } \partial\Omega_\varepsilon,\ s\geqslant 2.$$

Using mathematical induction and estimates (6), (7) of lemma 1, we obtain (16) for N^0_α. Estimate (16) for N^1_α holds since $N^1_\alpha(\varepsilon^{-1}x)$ are periodic functions of $x_1,\ldots,x_n$ with period ε. Since $N^0_\alpha = N^1_\alpha + N^2_\alpha$, it follows that (16) is valid for N^2_α. The lemma is proved. □

In order to show that the functions N^2_α are of boundary layer type we shall need the following theorem.

Consider a function $\tau(x) \in C^1(\bar{\Omega})$, $\tau = 0$ in a neighbourhood of $\partial\Omega$, $\tau \geqslant 0$ in Ω, $|\nabla\tau| \leqslant M = \text{const}$. Suppose that ε is so small that there exists a subdomain Ω' of Ω, whose closure $\overline{\Omega'}$ is a sum of sets $\varepsilon\bar{Q} + \varepsilon z$, where z takes the values from some subset I of $\mathbb{Z}^n$. Suppose also that $\partial\Omega'$ belongs to the neighbourhood of $\partial\Omega$ where $\tau = 0$.

THEOREM 1. Let $U(x)$ be a weak solution of problem (2) in Ω_ε, $\Phi = 0$ on $\overline{\Omega'} \cap \partial\Omega_\varepsilon$, $f_j \equiv 0$, $j = 1,\dots,n$. Then

$$\varepsilon^{-2} \int_{\Omega_\varepsilon \cap \Omega'} |U|^2 \exp\left(\frac{\delta\tau}{\varepsilon}\right) dx + \int_{\Omega_\varepsilon} |\nabla U|^2 \exp\left(\frac{\delta\tau}{\varepsilon}\right) dx \leqslant$$

$$\leqslant K\left[\varepsilon^2 \int_{\Omega_\varepsilon \cap \Omega} f_0^2 \exp\left(\frac{\delta\tau}{\varepsilon}\right) dx + \int_{\Omega_\varepsilon} |\nabla U|^2 dx\right], \tag{17}$$

where K, δ are positive constants, independent of U, f_0, Φ, ε.

Proof. The solution $U(x)$ satisfies the following integral identity

$$\int_{\Omega_\varepsilon} \frac{\partial U}{\partial x_i} \frac{\partial v}{\partial x_i} dx = -\int_{\Omega_\varepsilon} f_0 v \, dx$$

for any $v \in H_0^1(\Omega_\varepsilon)$. Let us take $v = U(e^{\mu\tau} - 1)$, where μ is a parameter to be chosen later. Since $\tau = 0$ outside of Ω' we have

$$\int_{\Omega_\varepsilon} |\nabla U|^2 e^{\mu\tau} dx = -\int_{\Omega_\varepsilon} \frac{\partial U}{\partial x_i} \mu \frac{\partial \tau}{\partial x_i} U e^{\mu\tau} dx - \int_{\Omega_\varepsilon} f_0 U(e^{\mu\tau} - 1) dx + \int_{\Omega_\varepsilon} |\nabla U|^2 dx \leqslant$$

$$\leqslant c_0\mu \left(\int_{\Omega' \cap \Omega_\varepsilon} |\nabla U|^2 e^{\mu\tau} dx\right)^{1/2} \left(\int_{\Omega' \cap \Omega_\varepsilon} |U|^2 e^{\mu\tau} dx\right)^{1/2} +$$

$$+ \left(\int_{\Omega' \cap \Omega_\varepsilon} f_0^2 e^{\mu\tau} dx\right)^{1/2} \left(\int_{\Omega' \cap \Omega_\varepsilon} |U|^2 e^{\mu\tau} dx\right)^{1/2} + \int_{\Omega_\varepsilon} |\nabla U|^2 dx. \tag{18}$$

Let $\omega_z = \varepsilon(Q \setminus G^0) + \varepsilon z \subset \Omega' \cap \Omega_\varepsilon$, $z \in I$. Using the Friedrichs inequality (5) for $U e^{\mu\tau/2}$ in every ω_z and taking into account that the constant in this inequality is of order ε^2 we get

$$\int_{\omega_z} |U|^2 e^{\mu\tau} dx \leqslant c_1 \varepsilon^2 \int_{\omega_z} |\nabla (U e^{\frac{\mu\tau}{2}})|^2 dx \leqslant$$

$$\leqslant c_2 \varepsilon^2 \int_{\omega_z} |\nabla U|^2 e^{\mu\tau} dx + c_3 \varepsilon^2 \mu^2 \int_{\omega_z} |U|^2 e^{\mu\tau} dx. \tag{19}$$

Here and in what follows the constant c_j do not depend on ε and U. Let us set $\mu = \sigma/2\varepsilon\sqrt{c_3}$, where $\sigma \in (0,1)$ is a constant independent of ε and will be chosen later. It is evident that for such μ inequality (19) implies

$$\int_{\omega_z} |U|^2 e^{\mu\tau} dx \leqslant c_4 \varepsilon^2 \int_{\omega_z} |\nabla U|^2 e^{\mu\tau} dx. \tag{20}$$

Since $\Omega' \cap \Omega_\varepsilon$ is a sum of sets ω_z, $z \in I$, we obtain from (20) that

$$\int_{\Omega' \cap \Omega_\varepsilon} |U|^2 e_{\mu\tau} dx \leqslant c_5 \varepsilon^2 \int_{\Omega' \cap \Omega_\varepsilon} |\nabla U|^2 e^{\mu\tau} dx. \tag{21}$$

Inequalities (21) and (18) yield

$$\int_{\omega_\varepsilon} |\nabla U|^2 e^{\mu\tau} dx \leqslant \mu\varepsilon c_0 \sqrt{c_5} \int_{\Omega' \cap \Omega_\varepsilon} |\nabla U|^2 e^{\mu\tau} dx +$$

$$+ \varepsilon\sqrt{c_5} \left(\int_{\Omega' \cap \Omega_\varepsilon} f_0^2 e^{\mu\tau} dx \right)^{1/2} \left(\int_{\Omega' \cap \Omega_\varepsilon} |\nabla U|^2 e^{\mu\tau} dx \right)^{1/2} + \int_{\Omega_\varepsilon} |\nabla U|^2 dx \leqslant$$

$$\leqslant \frac{\sigma}{2} \frac{\sqrt{c_5}}{\sqrt{c_3}} c_0 \int_{\Omega' \cap \Omega_\varepsilon} |\nabla U|^2 e^{\mu\tau} dx + \frac{1}{4} \int_{\Omega' \cap \Omega_\varepsilon} |\nabla U|^2 e^{\mu\tau} dx +$$

$$+ c_6 \varepsilon^2 \int_{\Omega' \cap \Omega_\varepsilon} f_0^2 e^{\mu\tau} dx + \int_{\Omega_\varepsilon} |\nabla U|^2 dx.$$

Taking $\sigma = \min(1, \sqrt{c_3}/c_0\sqrt{c_5})$ we obtain from this estimate that

$$\int_{\Omega_\varepsilon} |\nabla U|^2 e^{\mu\tau} dx \leqslant c_7 \varepsilon^2 \int_{\Omega \cap \Omega_\varepsilon} f_0^2 e^{\mu\tau} dx + c_8 \int_{\Omega_\varepsilon} |\nabla U|^2 dx. \tag{22}$$

Inequality (17) follows from (22), (21). The theorem is proved. □

<u>LEMMA 3</u>. Functions $N^2_\alpha(\varepsilon^{-1}x)$ <u>are</u> of boundary layer type, i.e. for any subdomain Ω^0 of Ω such that $\overline{\Omega^0}\subset\Omega$ the following estimate is valid

$$||N^2_\alpha(\varepsilon^{-1}x)||_{1,\Omega^0\cap\Omega_\varepsilon} \leqslant c_\alpha \exp(-\gamma\varepsilon^{-1}), \tag{23}$$

where c_α and γ are positive constants independent of ε.

<u>Proof</u>. We assume ε so small that there exists a subdomain $\Omega'\subset\Omega$ with the following properties : $\Omega^0\subset\Omega'$ and the distance between Ω^0 and $\partial\Omega'$ is larger than a positive constant independent of ε, Ω' consists of the sets $\varepsilon\bar{Q}+\varepsilon z$, $z\in I$, where I is a subset of $\mathbb{Z}^n$.

Let us take a function $\tau(x)$ such that $\tau\in C^1(\bar{\Omega})$, $\tau\equiv 1$ on Ω^0, $\tau\equiv 0$ outside Ω'. By induction we shall prove that $N^2_\alpha(\varepsilon^{-1}x)$ for $\alpha=(\alpha_1,\dots,\alpha_s)$ satisfy the following inequalities

$$\varepsilon^{-2}\int_{\Omega_\varepsilon\cap\Omega'} |N^2_\alpha|^2\exp\left(\frac{\delta\tau}{\varepsilon}\right)dx + \int_{\Omega_\varepsilon}|\nabla_x N^2_\varepsilon|^2\exp\left(\frac{\delta\tau}{\varepsilon}\right)dx \leqslant c_\alpha\varepsilon^{-2}, \tag{24}$$

where the constants c_α,δ do not depend on ε. First we show that (24) is valid for the function N^2_0 which is a solution of the following boundary value problem

$$\Delta_x N^2_0 = 0 \text{ in } \Omega_\varepsilon,\quad N^2_0 = -N^1_0 \quad\text{on } \partial\Omega_\varepsilon\ .$$

Using estimate (17) of theorem 1 for N^2_0 we get

$$\varepsilon^{-2}\int_{\Omega_\varepsilon\cap\Omega'}|N^2_0|^2\exp\left(\frac{\delta\tau}{\varepsilon}\right)dx + \int_{\Omega_\varepsilon}|\nabla_x N^2_0|^2\exp\left(\frac{\delta\tau}{\varepsilon}\right)dx \leqslant K\int_{\Omega_\varepsilon}|\nabla_x N^2_0|^2 dx.$$

This inequality together with (16) yields (24) for N^2_0. Now let s be a positive integer. Suppose that (24) holds for all N^2_α with $\alpha=(\alpha_1,\dots,\alpha_\ell)$, $\ell\leqslant s-1$. Let us show that (24) holds for $N^2_{\alpha_1\dots\alpha_s}$ which is a solution of the problem

$$\Delta_x N^2_{\alpha_1\ldots\alpha_s} = -2\varepsilon^{-1}\frac{\partial}{\partial x_{\alpha_1}} N^2_{\alpha_2\ldots\alpha_s} - \varepsilon^{-2}\delta_{\alpha_1\alpha_2} N^2_{\alpha_3\ldots\alpha_s} \quad \text{in } \partial\Omega,$$

$$N^2_{\alpha_1\ldots\alpha_s} = -N^1_{\alpha_1\ldots\alpha_s} \quad \text{on } \partial\Omega_\varepsilon.$$

It follows from (17) that

$$\varepsilon^{-2}\int_{\Omega_\varepsilon\cap\Omega'} |N^2_{\alpha_1\ldots\alpha_s}|^2 \exp(\frac{\delta\tau}{\varepsilon})dx + \int_{\Omega_\varepsilon} |\nabla_x N^2_{\alpha_1\ldots\alpha_s}|^2 \exp(\frac{\delta\tau}{\varepsilon})dx \leqslant$$

$$\leqslant K_1\Big[\varepsilon^2\varepsilon^{-2}\int_{\Omega_\varepsilon\cap\Omega'} |\nabla_x N^2_{\alpha_2\ldots\alpha_s}|^2 \exp(\frac{\delta\tau}{\varepsilon})dx +$$

$$+ \varepsilon^2\varepsilon^{-4}\int_{\Omega_\varepsilon\cap\Omega'} |N^2_{\alpha_3\ldots\alpha_s}|^2 \exp(\frac{\delta\tau}{\varepsilon})dx + \int_{\Omega_\varepsilon} |\nabla_x N^2_{\alpha_1\ldots\alpha_s}|^2 dx\Big].$$

Applying (16) to the last integral and using the induction assumption to estimate the first two integrals on the right hand side of (25), we get inequality (24) for $N^2_{\alpha_1\ldots\alpha_s}$. Obviously estimates (24) imply (23), since $\tau \equiv 1$ on Ω^0. The lemma is proved. □

The next theorem gives an estimate for the remainder term of the asymptotic expansion (9) of the solution $u_\varepsilon(x)$ of problem (1) with $\Phi = 0$ on $\partial\Omega_\varepsilon$.

<u>THEOREM 2.</u> Let $u_\varepsilon(x)$ be a solution of problem (1) with $\Phi = 0$, $f \in C^{s+2}(\bar{\Omega})$ and

$$u^s_\varepsilon(x) = \sum_{\ell=0}^{s} \varepsilon^{\ell+2} \sum_{\langle\alpha\rangle=\ell} N^0_\alpha(\frac{x}{\varepsilon}) D^\alpha f(x),$$

$$v^s_\varepsilon(x) = \sum_{\ell=0}^{s} \varepsilon^{\ell+2} \sum_{\langle\alpha\rangle=\ell} N^1_\alpha(\frac{x}{\varepsilon}) D^\varepsilon f(x),$$

$N^0_\alpha(\xi) = N^1_\alpha(\xi) + N^2_\alpha(\xi)$, N^1_α, N^2_α are weak solutions of problems (14), (15) respectively. Then

$$||u_\varepsilon(x) - u^s_\varepsilon(x)||_{1,\Omega_\varepsilon} \leqslant c_1\varepsilon^{s+1}, \tag{26}$$

and for any subdomain Ω^0 such that $\bar{\Omega}^0 \subset \Omega$,

$$\|u_\varepsilon(x) - v_\varepsilon^s(x)\|_{1,\Omega_\varepsilon \cap \Omega^0} \leqslant c_2 \varepsilon^{s+1}, \tag{27}$$

where constants c_1, c_2 do not depend on ε.

Let Ω^1 be a subdomain of Ω and $u_\varepsilon(x)$ be a solution of problem (1) with $f \equiv 0$ and $\Phi \equiv \Phi_\varepsilon = 0$ on $\Omega^1 \cap \partial\Omega_\varepsilon$. Then for any subdomain Ω^2, $\overline{\Omega^2} \subset \Omega^1$, there exist positive constants c, δ_1 independent of ε such that

$$\|u_\varepsilon\|_{1,\Omega^2 \cap \Omega_\varepsilon} \leqslant c\|\Phi_\varepsilon\|_{1,\Omega_\varepsilon} \exp(-\frac{\delta_1}{\varepsilon}). \tag{28}$$

<u>Proof</u>. Estimate (28) follows from theorem 1 and lemma 1. Let us prove estimate (26). Applying the Laplace operator to $N_\alpha^0(\frac{x}{\varepsilon}) D^\alpha f(x)$ we obtain

$$\Delta_x (N_\alpha^0(\frac{x}{\varepsilon}) D^\alpha f(x)) = \varepsilon^{-2} \frac{\partial^2 N_\alpha^0}{\partial \xi_i \partial \xi_i} D^\alpha f + 2\varepsilon^{-1} \frac{\partial N_\alpha^0}{\partial \xi_i} \frac{\partial D^\alpha f}{\partial x_i} + N_\alpha^0 D^\alpha \frac{\partial^2 f}{\partial x_i \partial x_i}.$$

Therefore

$$\Delta_x u_\varepsilon^s(x) = \sum_{m=0}^{s} \varepsilon^m \sum_{\alpha_1, \dots, \alpha_m = 1}^{n} \Delta_\xi N^0_{\alpha_1 \dots \alpha_m} D^{\alpha_1 \dots \alpha_m} f +$$

$$+ \sum_{m=1}^{s+1} \varepsilon^m \sum_{\alpha_1, \dots, \alpha_m = 1}^{n} 2 \frac{\partial N^0_{\alpha_2 \dots \alpha_m}}{\partial \xi_{\alpha_1}} D^{\alpha_1 \dots \alpha_m} f +$$

$$+ \sum_{m=2}^{s+2} \varepsilon^m \sum_{\alpha_1, \dots, \alpha_m = 1}^{n} \delta_{\alpha_1 \alpha_2} N^0_{\alpha_3 \dots \alpha_m} D^{\alpha_1 \dots \alpha_m} f.$$

Taking into account that N_α^0 satisfy (11)-(13) we see that $u_\varepsilon^s(x)$ is a weak solution of the following boundary problem $\Delta u_\varepsilon^s(x) = f(x) + f_s^\varepsilon(x)$ in Ω_ε, $u_\varepsilon^s = 0$ on $\partial\Omega_\varepsilon$, where

$$f_s^\varepsilon = \varepsilon^{s+1} \sum_{\alpha_1, \dots, \alpha_{s+1} = 1}^{n} 2 \frac{\partial N^0_{\alpha_2 \dots \alpha_{s+1}}}{\partial \xi_{\alpha_1}} D^{\alpha_1 \dots \alpha_{s+1}} f +$$

$$+\ \varepsilon^{s+1} \sum_{\alpha_1,\dots,\alpha_{s+1}=1}^{n} \delta_{\alpha_1\alpha_2} N^0_{\alpha_3\dots\alpha_{s+1}} D^{\alpha_1\dots\alpha_{s+1}} f\ +$$

$$+\ \varepsilon^{s+2} \sum_{\alpha_1,\dots,\alpha_{s+2}=1}^{n} \delta_{\alpha_1\alpha_2} N^0_{\alpha_3\dots\alpha_{s+2}} D^{\alpha_1\dots\alpha_{s+2}} f.$$

Estimate (26) follows from lemma 1 and the inequalities (16) of lemma 2. Estimate (27) holds by virtue of (26) and (23). The theorem is proved.

2. The biharmonic equation

In this section we obtain the asymptotic expansion for the solution $u_\varepsilon(x)$ of the Dirichlet problem in Ω_ε for the biharmonic equation

$$\Delta^2 u_\varepsilon(x) = f(x) \text{ in } \Omega_\varepsilon,\ u_\varepsilon = \Phi,\ \frac{\partial u_\varepsilon}{\partial \nu} = \frac{\partial \Phi}{\partial \nu} \text{ on } \partial\Omega_\varepsilon. \tag{29}$$

Denote by $H^2(\omega)$ the completion of the space $C^2(\bar{\omega})$ in the norm

$$\|u\|_{2,\omega} = \Big(\int_\omega \sum_{|\alpha|\leq 2} |\partial^\alpha u|^2 dx\Big)^{1/2},\ \alpha = (\alpha_1,\dots,\alpha_n),\ |\alpha| = \alpha_1+\dots+\alpha_n,$$

$$\partial^\alpha u = \partial^{|\alpha|} u/\partial x_1^{\alpha_1}\dots\partial x_n^{\alpha_n}.$$

As usual we denote by $H_0^2(\omega)$ the completion in the norm $\|u\|_{2,\omega}$ of the subspace of $C^2(\bar{\omega})$ formed by functions $u(x)$ vanishing in a neighbourhood of $\partial\omega$.

The set G^0 is supposed to be the same as in section 1, i.e. such that for any $v \in C^1(Q \setminus G^0)$, $v = 0$ in a neighbourhood of G^0, the Friedrichs inequality (5) is valid. It is easy to see that in this case for any $v \in H_0^2(\Omega_\varepsilon)$ we have

$$\varepsilon^{-4} \int_{\Omega_\varepsilon} |v|^2 dx + \varepsilon^{-2} \int_{\Omega_\varepsilon} |\nabla_x v|^2 dx \leq M_1 \int_{\Omega_\varepsilon} E_2(v) dx, \tag{30}$$

where $E_2(v) \equiv \dfrac{\partial^2 v}{\partial x_i \partial x_j} \dfrac{\partial^2 v}{\partial x_i \partial x_j}$, M_1 is a constant independent of ε.

Let $\Phi \in H^2(\Omega_\varepsilon)$, $f_0, f_i \in L_2(\Omega_\varepsilon)$, $i=1,\dots,n$. We say that $U(x)$ is a weak solution of the problem

$$\Delta^2 U(x) = f_0(x) + \frac{\partial f_i(x)}{\partial x_i} \quad \text{in } \Omega_\varepsilon,$$

$$U = \Phi, \ \frac{\partial U}{\partial \nu} = \frac{\partial \Phi}{\partial \nu} \quad \text{on } \partial\Omega_\varepsilon , \tag{31}$$

if the function $W \equiv U-\Phi$ belongs to $H_0^2(\Omega_\varepsilon)$ and satisfies the integral identity

$$\int_{\Omega_\varepsilon} \frac{\partial^2 W}{\partial x_i \partial x_j} \frac{\partial^2 v}{\partial x_i \partial x_j} dx = - \int_{\Omega_\varepsilon} \frac{\partial^2 \Phi}{\partial x_i \partial x_j} \frac{\partial^2 v}{\partial x_i \partial x_j} dx + \int_{\Omega_\varepsilon} (f_0 v - f_i \frac{\partial v}{\partial x_i}) dx \tag{32}$$

for any $v \in H_0^2(\Omega_\varepsilon)$.

Denote by $\hat{H}_0^2(Q \setminus G^0)$ the completion with respect to the norm $\|v\|_{2,Q\setminus G^0}$ of the functions $v(\xi)$ such that $v \in C^2(\mathbb{R}^N \setminus G_1)$, $v = 0$ in a neighbourhood of ∂G_1 and v is 1-periodic in ξ.

We say that $w(\xi)$ is 1-periodic in ξ weak solution of the problem

$$\Delta_\xi^2 w(\xi) = F_0(\xi) + \frac{\partial F_j(\xi)}{\partial \xi_j} \quad \text{in } \mathbb{R}^n \setminus G_1,$$

$$w = \frac{\partial w}{\partial \nu} = 0 \text{ on } \partial G_1, \tag{33}$$

where $F_0, F_j \in L_2(Q \setminus G^0)$, $j=1,\dots,n$, and are 1-periodic in ξ, if $w \in \hat{H}_0^2(Q \setminus G^0)$ and satisfies the integral identity

$$\int_{Q\setminus G^0} \frac{\partial^2 w}{\partial \xi_i \partial \xi_j} \frac{\partial^2 v}{\partial \xi_i \partial \xi_j} d\xi = \int_{Q\setminus G^0} (F_0 v - \frac{\partial v}{\partial \xi_j} F_j) d\xi$$

for any $v \in \hat{H}_0^2(Q \setminus G^0)$.

<u>LEMMA 4</u>. A weak solution $U(x)$ of problem (31) exists and satisfies the following inequalities

$$\int_{\Omega_\varepsilon} E_2(U) dx \leq K_1 \int_{\Omega_\varepsilon} [\varepsilon^4 |f_0|^2 + \varepsilon^2 f_i f_i + E^2(\Phi)] dx, \tag{34}$$

$$\int_{\Omega_\varepsilon} |\nabla U|^2 dx \leqslant K_2 \int_{\Omega_\varepsilon} [\varepsilon^6 |f_0|^2 + \varepsilon^4 f_i f_i + \varepsilon^2 E_2(\Phi) + |\nabla\Phi|^2] dx, \tag{35}$$

$$\int_{\Omega_\varepsilon} |U|^2 dx \leqslant K_3 \int_{\Omega_\varepsilon} [\varepsilon^8 |f_0|^2 + \varepsilon^6 f_i f_i + \varepsilon^4 E_2(\Phi) + |\Phi|^2] dx, \tag{36}$$

where constants K_1, K_2, K_3 do not depend on ε, f_0, f_i, Φ. □

The proof of this lemma is quite similar to that of lemma 1, and is based on the use of the Friedrichs inequality (30).

Let $f \in C^{s+4}(\bar{\Omega})$. We seek an approximate solution of problem (29) with $\Phi = 0$ in the form

$$u_\varepsilon^s(x) = \sum_{\ell=0}^{s} \varepsilon^{\ell+4} \sum_{\langle\alpha\rangle=\ell} N_\alpha^0(\xi) D^\alpha f(x), \quad \xi = \varepsilon^{-1} x, \tag{37}$$

where $N_\alpha^0(\xi)$ are functions to be determined.

It is easy to see that

$$\begin{aligned}
\Delta_x^2 (N_\alpha^0(\tfrac{x}{\varepsilon}) D^\alpha f(x)) &= \varepsilon^{-4} \Delta_\xi^2 N_\alpha^0(\xi) D^\alpha f(x) + 2\varepsilon^{-3} \frac{\partial}{\partial \xi_j} \Delta_\xi N_\alpha^0(\xi) \frac{\partial D^\alpha f(x)}{\partial x_j} + \\
&+ \varepsilon^{-2} \Delta_\xi N_\alpha^0(\xi) f(x) + 2\varepsilon^{-3} \Delta_\xi \frac{\partial N_\alpha^0(\xi)}{\partial \xi_i} D^\alpha \frac{\partial f(x)}{\partial x_i} + 4\varepsilon^{-2} \frac{\partial^2 N_\alpha^0(\xi)}{\partial \xi_i \partial \xi_j} D^\alpha \frac{\partial^2 f(x)}{\partial x_i \partial x_j} + \\
&+ 2\varepsilon^{-1} \frac{\partial N_\alpha^0(\xi)}{\partial \xi_i} D^\alpha \frac{\partial}{\partial x_i} \Delta_x f(x) + \varepsilon^{-2} \Delta_\xi N_\alpha^0(\xi) D^\alpha \frac{\partial^2 f(x)}{\partial x_i \partial x_i} + \\
&+ 2\varepsilon^{-1} \frac{\partial N_\alpha^0(\xi)}{\partial \xi_j} D^\alpha \frac{\partial^3 f}{\partial x_j \partial x_i \partial x_i} + N_\alpha^0(\xi) D^\alpha \frac{\partial^2 \Delta f(x)}{\partial x_i \partial x_i} .
\end{aligned} \tag{38}$$

Therefore from (37), (38) we have

$$\Delta^2 u_\varepsilon^s(x) = \sum_{m=0}^{s} \varepsilon^m \sum_{\alpha_1,\ldots,\alpha_m=1}^{n} \Delta_\xi^2 N_{\alpha_1\ldots\alpha_m}^0 D^{\alpha_1\ldots\alpha_m} f +$$

$$+ \sum_{m=1}^{s+1} \varepsilon^m \sum_{\alpha_1,\ldots,\alpha_m=1}^{n} 4 \frac{\partial}{\partial \xi_{\alpha_1}} \Delta_\xi N_{\alpha_2\ldots\alpha_m}^0 D^{\alpha_1\ldots\alpha_m} f +$$

$$+ \sum_{m=2}^{s+2} \varepsilon^m \sum_{\alpha_1,\dots,\alpha_m=1}^{n} \left(2\delta_{\alpha_1\alpha_2} \Delta_\xi N^0_{\alpha_3\dots\alpha_m} + 4 \frac{\partial^2 N^0_{\alpha_3\dots\alpha_m}}{\partial\xi_{\alpha_1}\partial\xi_{\alpha_2}}\right) D^{\alpha_1\dots\alpha_m} f +$$

$$+ \sum_{m=3}^{s+3} \varepsilon^m \sum_{\alpha_1,\dots,\alpha_m=1}^{n} 4\delta_{\alpha_1\alpha_2} \frac{\partial}{\partial\xi_{\alpha_3}} N^0_{\alpha_4\dots\alpha_m} D^{\alpha_1\dots\alpha_m} f +$$

$$+ \sum_{m=4}^{s+4} \varepsilon^m \sum_{\alpha_1,\dots,\alpha_m=1}^{n} \delta_{\alpha_1\alpha_2}\delta_{\alpha_3\alpha_4} N^0_{\alpha_5\dots\alpha_m} D^{\alpha_1\dots\alpha_m} f. \tag{39}$$

Let us define $N^0_\alpha(\xi)$ as weak solutions of the following boundary value problems

$$\Delta^2_\xi N^0_0 = 1 \text{ in } \varepsilon^{-1}\Omega_\varepsilon;\ N^0_0 = \frac{\partial N^0_0}{\partial\nu} = 0 \text{ on } \partial(\varepsilon^{-1}\Omega_\varepsilon),$$

$$\Delta^2_\xi N^0_{\alpha_1} = -4 \frac{\partial}{\partial\xi_{\alpha_1}}\Delta_\xi N^0_0 \text{ in } \varepsilon^{-1}\Omega_\varepsilon;\ N^0_{\alpha_1} = \frac{\partial N^0_{\alpha_1}}{\partial\nu} = 0 \text{ on } \partial(\varepsilon^{-1}\Omega_\varepsilon),$$

$$\Delta^2_\xi N^0_{\alpha_1\alpha_2} = -4 \frac{\partial}{\partial\xi_{\alpha_1}} \Delta_\xi N^0_{\alpha_2} - 2\delta_{\alpha_1\alpha_2}\Delta_\xi N^0_0 - 4 \frac{\partial^2 N^0_0}{\partial\xi_{\alpha_1}\partial\xi_{\alpha_2}} \text{ in } \varepsilon^{-1}\Omega_\varepsilon,$$

$$N^0_{\alpha_1\alpha_2} = \frac{\partial N^0_{\alpha_1\alpha_2}}{\partial\nu} = 0 \text{ on } \partial(\varepsilon^{-1}\Omega_\varepsilon),$$

$$\Delta^2_\xi N^0_{\alpha_1\alpha_2\alpha_3} = -4 \frac{\partial}{\partial\xi_{\alpha_1}} \Delta_\xi N^0_{\alpha_2\alpha_3} - 2\delta_{\alpha_1\alpha_2}\Delta_\xi N^0_{\alpha_3} - 4 \frac{\partial^2 N^0_{\alpha_3}}{\partial\xi_{\alpha_1}\partial\xi_{\alpha_2}} - 4\delta_{\alpha_1\alpha_2} \frac{\partial N^0_0}{\partial\xi_{\alpha_3}}$$

$$\text{in } \varepsilon^{-1}\Omega_\varepsilon,\ N^0_{\alpha_1\alpha_2\alpha_3} = \frac{\partial N^0_{\alpha_1\alpha_2\alpha_3}}{\partial\nu} = 0 \text{ on } \partial(\varepsilon^{-1}\Omega_\varepsilon),$$

$$\Delta^2_\xi N^0_{\alpha_1\dots\alpha_m} = -4 \frac{\partial}{\partial\xi_{\alpha_1}} \Delta_\xi N^0_{\alpha_2\dots\alpha_m} - 2\delta_{\alpha_1\alpha_2}\Delta_\xi N^0_{\alpha_3\dots\alpha_m} -$$

$$-4 \frac{\partial^2 N^0_{\alpha_3\dots\alpha_m}}{\partial\xi_{\alpha_1}\partial\xi_{\alpha_2}} - 4\delta_{\alpha_1\alpha_2} \frac{\partial N^0_{\alpha_4\dots\alpha_m}}{\partial\xi_{\alpha_3}} - \delta_{\alpha_1\alpha_2}\delta_{\alpha_3\alpha_4} N_{\alpha_5\dots\alpha_m} \text{ in } \varepsilon^{-1}\Omega_\varepsilon,$$

$$N^0_{\alpha_1\ldots\alpha_m} = \frac{\partial N^0_{\alpha_1\ldots\alpha_m}}{\partial\nu} = 0 \text{ on } \partial(\varepsilon^{-1}\Omega_\varepsilon),\ m\geqslant 4. \qquad (40)$$

It is easy to prove by induction that N^0_α exist, $N^0_\alpha(\varepsilon^{-1}x) \in H^2_0(\Omega_\varepsilon)$.

Let us show now that $N^0_\alpha(\xi) = N^1_\alpha(\xi) + N^2_\alpha(\xi)$ where $N^1_\alpha(\xi)$ are 1-periodic in ξ functions which belong to $\hat{H}^2_0(Q\setminus G^0)$, and $N^2_\xi(\varepsilon^{-1}x)$ are of boundary layer type in Ω_ε. Set

$$T^1_0 \equiv 1,\ T^2_0 \equiv 0,\ T^j_{\alpha_1} \equiv -4\frac{\partial}{\partial\xi_{\alpha_1}}\Delta_\xi N^j_0,$$

$$T^j_{\alpha_1\alpha_2} \equiv -4\frac{\partial}{\partial\xi_{\alpha_1}}\Delta_\xi N^j_{\alpha_2} - 2\delta_{\alpha_1\alpha_2}\Delta_\xi N^j_0 - 4\frac{\partial^2 N^j_0}{\partial\xi_{\alpha_1}\partial\xi_{\alpha_2}},$$

$$T^j_{\alpha_1\alpha_2\alpha_3} \equiv -4\frac{\partial}{\partial\xi_{\alpha_1}}\Delta_\xi N^j_{\alpha_2\alpha_3} - 2\delta_{\alpha_1\alpha_2}\Delta_\xi N^j_{\alpha_3} - 4\frac{\partial^2 N^j_{\alpha_3}}{\partial\xi_{\alpha_1}\partial\xi_{\alpha_2}} - 4\delta_{\alpha_1\alpha_2}\frac{\partial N^j_0}{\partial\xi_{\alpha_3}},$$

$$T^j_{\alpha_1\ldots\alpha_m} \equiv -4\frac{\partial}{\partial\xi_{\alpha_1}}\Delta_\xi N^j_{\alpha_2\ldots\alpha_m} - 2\delta_{\alpha_1\alpha_2}\Delta_\xi N^j_{\alpha_3\ldots\alpha_m} - 4\frac{\partial^2 N^j_{\alpha_3\ldots\alpha_m}}{\partial\xi_{\alpha_1}\partial\xi_{\alpha_2}} -$$

$$-4\delta_{\alpha_1\alpha_2}\frac{\partial N^j_{\alpha_4\ldots\alpha_m}}{\partial\xi_{\alpha_3}} - \delta_{\alpha_1\alpha_2}\delta_{\alpha_3\alpha_4}N^j_{\alpha_5\ldots\alpha_m},\ j = 1,2,\ m\geqslant 4.$$

We define N^1_α as 1-periodic in ξ weak solutions of the problems

$$\Delta^2_\xi N^1_\alpha(\xi) = T^1_\alpha(\xi) \text{ in } \mathbb{R}^n\setminus G_1,$$

$$N^1_\alpha = \frac{\partial N^1_\alpha}{\partial\nu} = 0 \text{ on } \partial G_1. \qquad (41)$$

It follows from the Lax-Milgram lemma and the Friedrichs inequality that N^1_α exist.

In $\varepsilon^{-1}\Omega_\varepsilon$ we define functions $N^2_\alpha(\xi)$ as weak solutions of the problems

$$\Delta_\xi^2 N_\alpha^2(\xi) = T_\alpha^2(\xi) \text{ in } \varepsilon^{-1}\Omega_\varepsilon, \tag{42}$$

$$N_\alpha^2(\xi) = -N_\alpha^1(\xi), \quad \frac{\partial N_\alpha^2(\xi)}{\partial\nu} = -\frac{\partial N_\alpha^1(\xi)}{\partial\nu} \quad \text{on } \partial(\varepsilon^{-1}\Omega_\varepsilon).$$

Existence of N_α^2 follows from lemma 4.

LEMMA 5. Functions $N_\alpha^j(\varepsilon^{-1}x)$ satisfy the following inequalities

$$\varepsilon^4\int_{\Omega_\varepsilon} E_2(N_\alpha^j)\,dx + \varepsilon^2\int_{\Omega_\varepsilon} |\nabla_x N_\alpha^j|^2 dx + \int_{\Omega_\varepsilon} |N^j|^2 dx \leqslant M_\alpha, \quad j = 0,1,2, \tag{43}$$

where constants M_α do not depend on ε. □

The proof of this lemma is based on lemma 4 and is quite similar to the proof of lemma 2 of section 1.

Let $\tau(x)$ be a function in $C^2(\bar\Omega)$, $\tau = 0$ in a neighbourhood of $\partial\Omega$, $\tau \geqslant 0$ in Ω, $\|\tau\|_{C^2(\bar\Omega)} \leqslant M$. Consider Ω', the subdomain of Ω, defined just before theorem 1.

THEOREM 3. Suppose that $U(x)$ is a weak solution of problem (31) in Ω_ε and $\Phi = \frac{\partial\Phi}{\partial\nu} = 0$ on $\bar\Omega' \cap \partial\Omega_\varepsilon$. Then

$$\varepsilon^{-4}\int_{\Omega_\varepsilon\cap\Omega'} |U|^2 \exp\left(\frac{\delta\tau}{\varepsilon}\right) dx + \varepsilon^{-2}\int_{\Omega_\varepsilon\cap\Omega'} |\nabla_x U|^2 \exp\left(\frac{\delta\tau}{\varepsilon}\right) dx + \int_{\Omega_\varepsilon} E_2(U)\exp\left(\frac{\delta\tau}{\varepsilon}\right) dx \leqslant$$

$$\leqslant K_0\left(\varepsilon^4\int_{\Omega_\varepsilon\cap\Omega'} f_0^2 \exp\left(\frac{\delta\tau}{\varepsilon}\right) dx + \varepsilon^2\int_{\Omega_\varepsilon\cap\Omega'} f_i f_i \exp\left(\frac{\delta\tau}{\varepsilon}\right) dx + \int_{\Omega_\varepsilon} E_2(U)\,dx\right), \tag{44}$$

where $K_0 > 0$, $\delta > 0$ are constants independent of ε.

Proof. The solution $U(x)$ satisfies the integral identity

$$\int_{\Omega_\varepsilon} \frac{\partial^2 U}{\partial x_i \partial x_j} \frac{\partial^2 v}{\partial x_i \partial x_j}\,dx = \int_{\Omega_\varepsilon} \left(f_0 v - \frac{\partial v}{\partial x_j} f_j\right) dx$$

for any $v \in H_0^2(\Omega_\varepsilon)$. Let us take $v = U(e^{\mu\tau}-1)$ where μ is a parameter to the chosen later. Then we have

$$\int_{\Omega_\varepsilon} \frac{\partial^2 U}{\partial x_i \partial x_j} \frac{\partial^2 U}{\partial x_i \partial x_j} e^{\mu\tau} dx = -2\mu \int_{\Omega_\varepsilon} \frac{\partial^2 U}{\partial x_i \partial x_j} \frac{\partial U}{\partial x_i} \frac{\partial \tau}{\partial x_j} e^{\mu\tau} dx -$$

$$-\mu \int_{\Omega_\varepsilon} \frac{\partial^2 U}{\partial x_i \partial x_j} U \frac{\partial^2 \tau}{\partial x_i \partial x_j} e^{\mu\tau} dx - \mu^2 \int_{\Omega_\varepsilon} \frac{\partial^2 U}{\partial x_i \partial x_j} U \frac{\partial \tau}{\partial x_i} \frac{\partial \tau}{\partial x_j} e^{\mu\tau} dx +$$

$$+ \int_{\Omega_\varepsilon} fU(e^{\mu\tau}-1)dx - \int_{\Omega_\varepsilon} f_i \frac{\partial U}{\partial x_i} (e^{\mu\tau}-1)dx - \int_{\Omega_\varepsilon} f_i U_\mu \frac{\partial \tau}{\partial x_i} e^{\mu\tau} dx +$$

$$+ \int_{\Omega_\varepsilon} \frac{\partial^2 U}{\partial x_i \partial x_j} \frac{\partial^2 U}{\partial x_i \partial x_j} dx. \tag{45}$$

Taking into account that $\tau \equiv 0$ outside of Ω' and using Hölder's inequality, we obtain from (45) that

$$\int_{\Omega_\varepsilon} E_2(U) e^{\mu\tau} dx \leqslant c_1 \mu \Big(\int_{\Omega_\varepsilon \cap \Omega'} E_2(U) e^{\mu\tau} dx \Big)^{1/2} \Big(\int_{\Omega_\varepsilon \cap \Omega'} |\nabla U|^2 e^{\mu\tau} dx \Big)^{1/2} +$$

$$+ c_2(\mu+\mu^2) \Big(\int_{\Omega_\varepsilon \cap \Omega'} E_2(U) e^{\mu\tau} dx \Big)^{1/2} \Big(\int_{\Omega_\varepsilon \cap \Omega'} |\nabla U|^2 e^{\mu\tau} dx \Big)^{1/2} +$$

$$+ c_3 \Big(\int_{\Omega_\varepsilon \cap \Omega'} f_0^2 (e^{\mu\tau}-1) dx \Big)^{1/2} \Big(\int_{\Omega_\varepsilon \cap \Omega'} |U|^2 e^{\mu\tau} dx \Big)^{1/2} +$$

$$+ c_4 \Big(\int_{\Omega_\varepsilon \cap \Omega'} f_i f_i (e^{\mu\tau}-1) dx \Big)^{1/2} \Big(\int_{\Omega_\varepsilon \cap \Omega'} |\nabla U|^2 e^{\mu\tau} dx \Big)^{1/2} +$$

$$+ c_5 \mu \Big(\int_{\Omega_\varepsilon \cap \Omega'} f_i f_i e^{\mu\tau} dx \Big)^{1/2} \Big(\int_{\Omega_\varepsilon \cap \Omega'} |U|^2 e^{\mu\tau} dx \Big)^{1/2} + \int_{\Omega_\varepsilon} E_2(U) dx. \tag{46}$$

In the same way as it is done in the proof of theorem 1 we get the inequality

$$\int_{\Omega_\varepsilon \cap \Omega'} |U|^2 e^{\mu\tau} dx \leqslant K_1 \varepsilon^2 \int_{\Omega_\varepsilon \cap \Omega'} |\nabla U|^2 e^{\mu\tau} dx, \quad K_1 = \text{const.}, \tag{47}$$

where $\mu = \sigma/K\varepsilon$, K is a constant independent of ε, $\sigma \in (0,1)$ and σ will be chosen later.

Since $U(x)$ can be approximated in the norm $H^2(\Omega_\varepsilon \cap \Omega')$ by functions vanishing in a neighbourhood of $\varepsilon G_1 \cap \bar{\Omega}$, it follows that inequality similar to (47) holds for the first derivatives of $U(x)$, i.e.

$$\int_{\Omega_\varepsilon \cap \Omega'} |\nabla U|^2 e^{\mu\tau} dx \leqslant K_2 \varepsilon^2 \int_{\Omega_\varepsilon \cap \Omega'} E_2(U) e^{\mu\tau} dx, \quad K_2 = \text{const.} \tag{48}$$

Estimates (47), (48) yield

$$\int_{\Omega_\varepsilon \cap \Omega'} |U|^2 e^{\mu\tau} dx \leqslant K_3 \varepsilon^4 \int_{\Omega_\varepsilon \cap \Omega'} E_2(U) e^{\mu\tau} dx, \quad K_3 = \text{const.} \tag{49}$$

From (46), (48), (49) we obtain

$$\begin{aligned}
\int_{\Omega_\varepsilon} E_2(U) e^{\mu\tau} dx \leqslant{} & c_6 \mu\varepsilon \int_{\Omega_\varepsilon \cap \Omega'} E_2(U) e^{\mu\tau} dx + c_7(\mu+\mu^2)\varepsilon^2 \int_{\Omega_\varepsilon \cap \Omega'} E_2(U) e^{\mu\tau} dx + \\
& + c_8 \varepsilon^2 \Big(\int_{\Omega_\varepsilon \cap \Omega'} f_0^2 e^{\mu\tau} dx\Big)^{1/2} \Big(\int_{\Omega_\varepsilon \cap \Omega'} E_2(U) e^{\mu\tau} dx\Big)^{1/2} + \\
& + c_9 \varepsilon \Big(\int_{\Omega_\varepsilon \cap \Omega'} f_i f_i e^{\mu\tau} dx\Big)^{1/2} \Big(\int_{\Omega_\varepsilon \cap \Omega'} E_2(U) e^{\mu\tau} dx\Big)^{1/2} + \\
& + c_{10} \mu\varepsilon^2 \Big(\int_{\Omega_\varepsilon \cap \Omega'} f_i f_i e^{\mu\tau} dx\Big)^{1/2} \Big(\int_{\Omega_\varepsilon \cap \Omega'} E_2(U) e^{\mu\tau} dx\Big)^{1/2} + \int_{\Omega_\varepsilon \cap \Omega'} E_2(U) dx \,.
\end{aligned} \tag{50}$$

where $\mu = \sigma/K\varepsilon$. If we chose σ sufficiently small but independent of ε, we get from (50) the following inequality

$$\begin{aligned}
\int_{\Omega_\varepsilon} E_2(U) e^{\mu\tau} dx \leqslant{} & M_1 \varepsilon^4 \int_{\Omega_\varepsilon \cap \Omega'} f_0^2 e^{\mu\tau} dx + M_2 \varepsilon^2 \int_{\Omega_\varepsilon \cap \Omega'} f_i f_i e^{\mu\tau} dx + \\
& + M_3 \int_{\Omega_\varepsilon} E_2(U) dx,
\end{aligned} \tag{51}$$

where constants M_1, M_2, M_3 do not depend on ε. Estimate (44) follows from (51), (48), (49). The theorem is proved. □

Let us now prove a theorem on the asymptotic expansion with respect to ε of

$u_\varepsilon(x)$ which is a solution of problem (29).

THEOREM 4. Let $u_\varepsilon(x)$ be a weak solution of problem (29) in Ω_ε with $\Phi \equiv 0$, $f \in C^{s+4}(\bar{\Omega})$, $u_\varepsilon^s(x)$ is given by the formula (37) and

$$v_\varepsilon^s(x) = \sum_{\ell=0}^{s} \varepsilon^{\ell+4} \sum_{<\alpha>=\ell} N_\alpha^\ell(\frac{x}{\varepsilon}) D^\alpha f(x),$$

where $N_\alpha^0 = N_\alpha^1 + N_\alpha^2$, N_α^1, N_α^2 are weak solutions of problems (41), (42) respectively. Then

$$\|u_\varepsilon(x) - v_\varepsilon^s(x)\|_{2,\Omega_\varepsilon} \leqslant c\varepsilon^{s+1} , \tag{52}$$

where C is a constant independent of ε. The functions N_α^2 are of boundary layer type, i.e. for any subdomain Ω^0 such that $\overline{\Omega^0} \subset \Omega$ the following estimate holds

$$\|N^2(\frac{x}{\varepsilon})\|_{2,\Omega^0 \cap \Omega_\varepsilon} \leqslant c_\alpha \exp(-\gamma\varepsilon^{-1}) , \tag{53}$$

moreover

$$\|u_\varepsilon(x) - v_\varepsilon^s(x)\|_{2,\Omega^0 \cap \Omega_\varepsilon} \leqslant M_1 \varepsilon^{s+1}, \tag{54}$$

where c_α, M_1, γ are positive constants independent of ε.

Let Ω^1 be a subdomain of Ω and $u_\varepsilon(x)$ be a solution of problem (29) with $f \equiv 0$ in Ω, $\Phi \equiv \Phi_\varepsilon$, $\Phi_\varepsilon = \frac{\partial \Phi_\varepsilon}{\partial \nu} = 0$ on $\Omega^1 \cap \partial\Omega_\varepsilon$. Then for any subdomain Ω^2, $\overline{\Omega^2} \subset \Omega^1$, there exist positive constants c, δ independent of ε and such that

$$\|u_\varepsilon\|_{2,\Omega^2 \cap \Omega_\varepsilon} \leqslant c\|\Phi_\varepsilon\|_{2,\Omega_\varepsilon} \exp(-\frac{\delta}{\varepsilon}) \tag{55}$$

Proof. Estimate (55) follows from theorem 3 and lemma 4, so we shall consider the case $\Phi = \frac{\partial \Phi}{\partial \nu} = 0$ on $\partial\Omega_\varepsilon$. It is easy to see from (39), (40) that $u_\varepsilon^s(x)$ is a weak solution of the following boundary value problem

$$\Delta^2 u_\varepsilon^s(x) = f+\varepsilon^{s+1} \sum_{\alpha_1,\dots,\alpha_{s+1}=1}^{n} 4\varepsilon^3 \frac{\partial}{\partial x_{\alpha_1}}\left[\Delta_x N_{\alpha_2\dots\alpha_{s+1}} D^{\alpha_1\dots\alpha_{s+1}} f \right]-$$

$$\varepsilon^{s+1} \sum_{\alpha_1,\dots,\alpha_{s+1}=1}^{n} 4\varepsilon^2 \Delta_x N_{\alpha_2\dots\alpha_{s+1}} \frac{\partial D^{\alpha_1\dots\alpha_{s+1}}}{\partial x_{\alpha_1}} f +$$

$$+ \sum_{m=s+1}^{s+2} \varepsilon^m \sum_{\alpha_1,\dots,\alpha_m=1}^{n} (2\delta_{\alpha_1\alpha_2} \varepsilon^2 \Delta_x N_{\alpha_3\dots\alpha_m} + 4\varepsilon^3 \frac{\partial^2 N_{\alpha_3\dots\alpha_m}}{\partial x_{\alpha_1}\partial x_{\alpha_2}}) D^{\alpha_1\dots\alpha_m} f+$$

$$+ \sum_{m=s+1}^{s+3} \varepsilon^m \sum_{\alpha_1,\dots,\alpha_m=1}^{n} 4\delta_{\alpha_1\alpha_2} \varepsilon \frac{\partial N_{\alpha_4\dots\alpha_m}}{\partial x_{\alpha_3}} D^{\alpha_1\dots\alpha_m} f +$$

$$+ \sum_{m=s+1}^{s+4} \varepsilon^m \sum_{\alpha_1,\dots,\alpha_m=1}^{n} \delta_{\alpha_1\alpha_2} \delta_{\alpha_3\alpha_4} N_{\alpha_5\dots\alpha_m} D^{\alpha_1\dots\alpha_m} f \text{ in } \Omega_\varepsilon,$$

$$u_\varepsilon^s(x) = \frac{\partial u_\varepsilon^s(x)}{\partial \nu} = 0 \text{ on } \partial\Omega_\varepsilon.$$

Therefore estimate (52) is valid due to lemma 4 and lemma 5. Let us prove estimate (53). The function $N^2_{\alpha_1\dots\alpha_m}(\varepsilon^{-1}x)$ is a weak solution of the boundary value problem

$$\Delta_x^2 N^2_{\alpha_1\dots\alpha_m} = -\varepsilon^{-1} 4 \frac{\partial}{\partial x_{\alpha_1}} \Delta_x N^2_{\alpha_2\dots\alpha_m} - \varepsilon^{-2} 2\delta_{\alpha_1\alpha_2} \Delta_x N^2_{\alpha_2\dots\alpha_m} -$$

$$- \varepsilon^{-2} 4 \frac{\partial^2 N^2_{\alpha_3\dots\alpha_m}}{\partial x_{\alpha_1}\partial x_{\alpha_2}} - \varepsilon^{-3} 4\delta_{\alpha_1\alpha_2} \frac{\partial N^2_{\alpha_4\dots\alpha_m}}{\partial x_{\alpha_3}} - \varepsilon^{-4} \delta_{\alpha_1\alpha_2} N^2_{\alpha_5\dots\alpha_m} \text{ in } \Omega_\varepsilon$$

$$N^2_{\alpha_1\dots\alpha_m} = -N^1_{\alpha_1\dots\alpha_m}, \quad \frac{\partial N^2_{\alpha_1\dots\alpha_m}}{\partial \nu} = - \frac{\partial N^1_{\alpha_1\dots\alpha_n}}{\partial \nu} \text{ on } \partial\Omega_\varepsilon.$$

Let Ω' be a subdomain of Ω defined just before theorem 1. Then using estimate (44) of theorem 3 for $N^2_{\alpha_1\dots\alpha_m}$, we find that

$$\varepsilon^{-4} \int_{\Omega_\varepsilon \cap \Omega'} |N^2_{\alpha_1 \ldots \alpha_m}|^2 \exp(\frac{\delta\tau}{\varepsilon}) dx + \varepsilon^{-2} \int_{\Omega_\varepsilon \cap \Omega'} |\nabla N^2_{\alpha_1 \ldots \alpha_m}|^2 \exp(\frac{\delta\tau}{\varepsilon}) dx +$$

$$+ \int_{\Omega_\varepsilon} E_2(N^2_{\alpha_1 \ldots \alpha_m}) \exp(\frac{\delta\tau}{\varepsilon}) dx \leq K_1 \Big(\varepsilon^4 \int_{\Omega_\varepsilon \cap \Omega'} \Big[\varepsilon^{-4} E_2(N^2_{\alpha_2 \ldots \alpha_m}) +$$

$$+ \varepsilon^{-4} E_2(N^2_{\alpha_3 \ldots \alpha_m}) + \varepsilon^{-6} |\nabla_x N^2_{\alpha_4 \ldots \alpha_m}|^2 + \varepsilon^{-8} |N^2_{\alpha_5 \ldots \alpha_m}|^2 \Big] \exp(\frac{\delta\tau}{\varepsilon}) dx +$$

$$+ \varepsilon^2 \int_{\Omega_\varepsilon \cap \Omega'} \varepsilon^{-2} E_2(N^2_{\alpha_2 \ldots \alpha_m}) \exp(\frac{\delta\tau}{\varepsilon}) dx + \int_{\Omega_\varepsilon} E_2(N^2_{\alpha_1 \ldots \alpha_m}) dx \Big) \leq$$

$$\leq K_2 \Big(\int_{\Omega_\varepsilon \cap \Omega'} \Big[E_2(N^2_{\alpha_2 \ldots \alpha_m}) + E_2(N^2_{\alpha_3 \ldots \alpha_m}) + \varepsilon^{-2} |\nabla_x N^2_{\alpha_4 \ldots \alpha_m}|^2 +$$

$$+ \varepsilon^{-4} |N^2_{\alpha_5 \ldots \alpha_m}|^2 \Big] \exp(\frac{\delta\tau}{\varepsilon}) dx + \varepsilon^{-4} \Big) ,$$

where constant K_2 does not depend on ε. By virtue of this inequality it is easy to show by induction that the functions $N^2_{\alpha_1 \ldots \alpha_m}$ satisfy the inequality

$$\varepsilon^{-4} \int_{\Omega_\varepsilon \cap \Omega'} |N^2_{\alpha_1 \ldots \alpha_m}|^2 \exp(\frac{\delta\tau}{\varepsilon}) dx + \varepsilon^{-2} \int_{\Omega_\varepsilon \cap \Omega'} |\nabla_x N^2_{\alpha_1 \ldots \alpha_m}|^2 \exp(\frac{\delta\tau}{\varepsilon}) dx +$$

$$+ \int_{\Omega_\varepsilon} E_2(N^2_{\alpha_1 \ldots \alpha_m}) \exp(\frac{\delta\tau}{\varepsilon}) dx \leq K_{\alpha_1 \ldots \alpha_m} \varepsilon^{-4} ,$$

where constants $K_{\alpha_1 \ldots \alpha_m}$ do not depend on ε. Therefore taking into account that $\tau \equiv 1$ on Ω^0 we obtain estimate (53). Estimate (54) follows from (52), (53). The theorem is proved.

3. The system of linear elasticity

Consider in Ω_ε the boundary value problem for the linear elasticity system

$$L_\varepsilon(u_\varepsilon) \equiv \frac{\partial}{\partial x_h} (C^{hk}(\frac{x}{\varepsilon}) \frac{\partial u_\varepsilon}{\partial x_k}) = f(x) \text{ in } \Omega_\varepsilon \tag{56}$$

$u_\varepsilon = \Phi$ on $\partial\Omega_\varepsilon$,

where $C^{hk}(\xi)$ are $(n \times n)$ matrices whose elements $c^{hk}_{ij}(\xi)$ are bounded measurable functions, 1-periodic in ξ, $u_\varepsilon = (u^\varepsilon_1, \ldots, u^\varepsilon_n)^*$, $f = (f_1, \ldots, f_n)^*$ are vector columns. The coefficients c^{hk}_{ij} in (56) are assumed to satisfy the conditions usual for the elasticity system

$$c^{hk}_{ij}(\xi) = c^{kh}_{ji}(\xi) = c^{ik}_{hj}(\xi), \quad \xi \in Q \setminus G^0, \tag{57}$$

$$A_1 \eta^h_i \eta^h_i \leqslant c^{hk}(\xi)\eta^h \eta^k \leqslant A_2 \eta^h_i \eta^h_i, \quad A_1, A_2 = \text{const}, \quad \xi \in Q \setminus G^0,$$

for every symmetric $(n \times n)$ matrix $\eta = \{\eta^s_t\}$.

As in sections 1 and 2 a weak solution of problem (56) is a vector valued function u_ε with components in $H^1(\Omega_\varepsilon)$ such that the components of $u_\varepsilon - \Phi$ belong to $H^1_0(\Omega_\varepsilon)$ and satisfy the corresponding integral identity.

In this section we assume the set G^0 to be such that for every vector valued function v in $C^1(Q)$ vanishing in a neighbourhood of G^0, the Korn inequality is valid

$$\int_{Q \setminus G^0} (|v|^2 + |\nabla_\xi v|^2)\, d\xi \leqslant M \int_{Q \setminus G^0} E_\xi(v)\, d\xi, \tag{58}$$

where M is a constant independent of v, $E_\xi(v) \equiv \sum_{i,j=1}^{n} \left(\frac{\partial v_i}{\partial x_j} + \frac{\partial v_j}{\partial x_i} \right)^2$.

A sufficient condition for (58) is given by

<u>LEMMA 6</u>. Suppose that there exists a $(n-1)$ dimensional surface Γ of class C^1 belonging to G^0. Then inequality (58) holds for any $v \in C^1(Q)$ vanishing in a neighbourhood of G^0.

<u>Proof</u>. It follows from the results of [14] that for any $w \in H^1(Q)$ the Korn inequality of the form

$$\int_Q (|w|^2 + |\nabla w|^2)\, d\xi \leqslant K \int_Q [|w|^2 + E_\xi(w)]\, d\xi \tag{59}$$

is satisfied. Suppose that the assertion of lemma 6 is not true. Then there exists a sequence of vector valued functions $v^m \in C^1(Q)$ such that each v^m vanishes in a neighbourhood of G^0 and satisfies the inequality

$$\int_{Q\setminus G^0} (|v^m|^2 + |\nabla_\xi v^m|^2)\, d\xi \geq m \int_{Q\setminus G^0} E_\xi(v^m)\, d\xi, \quad m = 1,2,\ldots . \tag{60}$$

We can assume that $\|v^m\|_{1,Q\setminus G^0} = 1$. Then

$$\int_{Q\setminus G^0} E_\xi(v^m)\, d\xi \leq \frac{1}{m}, \quad m = 1,2,\ldots \tag{61}$$

It follows from the compactness of the imbedding $H^1(Q) \subset L_2(Q)$ and from the weak compactness of a unit ball in $H^1(Q)$ that there exists a subsequence v^{m_k} such that $v^{m_k} \to V$ in $L_2(Q)$, $v^{m_k} \to V$ weakly in $H^1(Q)$ as $m \to \infty$. Using inequality (59) to $v^{m_k} - v^{m_r}$, we find that

$$\int_Q (|v^{m_k} - v^{m_r}|^2 + |\nabla_\xi (v^{m_k} - v^{m_r})|^2)\, d\xi \leq K \int_Q (|v^{m_k} - v^{m_r}|^2 + E_\xi(v^{m_k} - v^{m_r}))\, d\xi.$$

Therefore $v^{m_k} \to V$ strongly in $H^1(Q)$ due to (61) and the strong convergence of v^{m_k} in $L_2(Q)$. Thus we have $\int_Q E_\xi(V)\, d\xi = 0$ which means (see [15]) that $V(\xi)$ is a rigid displacement, i.e. $V = A + B\xi$, where A is a constant vector and B is a skew-symmetrical matrix with constant elements. It follows from the imbedding theorems that $V = 0$ on Γ. One can easily prove that $V \equiv 0$ using the explicit form of V. Therefore $V \equiv 0$ and $\|V\|_{1,Q\setminus G^0} = 1$.

This contradiction shows that inequality (58) is valid. The lemma is proved.

LEMMA 7. Let $U(x) \in H^1(\Omega_\varepsilon)$ be a weak solution of problem

$$L_\varepsilon(U) = f^0(x) + \frac{\partial f^j}{\partial x_j} \quad \text{in } \Omega_\varepsilon,$$

$$U = \Phi \quad \text{on } \partial\Omega_\varepsilon, \tag{62}$$

where $f^0, f^j \in L_2(\Omega_\varepsilon)$, $j = 1,\ldots,n$, $\Phi \in H^1(\Omega_\varepsilon)$. Then

$$\int_{\Omega_\varepsilon} |\nabla U|^2 dx \leqslant c(\varepsilon^2 \int_{\Omega_\varepsilon} |f^0|^2 dx + \int_{\Omega_\varepsilon} (f^i, f^i) dx + \int_{\Omega_\varepsilon} |\nabla\Phi|^2 dx), \tag{63}$$

$$\int_{\Omega_\varepsilon} |U|^2 dx \leqslant c_1(\varepsilon^4 \int_{\Omega_\varepsilon} |f^0|^2 dx + \varepsilon^2 \int_{\Omega_\varepsilon} (f^i, f^i) dx + \varepsilon^2 \int_{\Omega_\varepsilon} |\nabla\Phi|^2 dx + \int_{\Omega_\varepsilon} |\Phi|^2 dx), \tag{64}$$

where constants c, c_1 do not depend on ε, $(u,v) = u_i v_i$.

Proof. It follows from lemma 6 that for any vector valued function $w \in H_0^1(\Omega_\varepsilon)$ we have

$$\int_{\Omega_\varepsilon} |w|^2 dx \leqslant M\varepsilon^2 \int_{\Omega_\varepsilon} E(w) dx, \quad \int_{\Omega_\varepsilon} |\nabla w|^2 dx \leqslant M \int_{\Omega_\varepsilon} E(w) dx, \tag{65}$$

where $E(w) \equiv \sum_{i,j=1}^{n} \left(\frac{\partial w_i}{\partial x_j} + \frac{\partial w_j}{\partial x_i}\right)^2$. The integral identity for $w = U - \Phi$ yields the inequality

$$\int_{\Omega_\varepsilon} c_{ij}^{hk} \frac{\partial w_j}{\partial x_h} \frac{\partial w_i}{\partial x_h} dx \leqslant \left|\int_{\Omega_\varepsilon} (f^0, w) dx\right| + \left|\int_{\Omega_\varepsilon} (f^i, \frac{\partial w}{\partial x_i}) dx\right| + \left|\int_{\Omega_\varepsilon} c_{ij}^{hk} \frac{\partial w_j}{\partial x_k} \frac{\partial \Phi_i}{\partial x_h} dx\right|$$

By virtue of (57) we have

$$\int_{\Omega_\varepsilon} E(w) dx \leqslant K \int_{\Omega_\varepsilon} c_{ij}^{hk} \frac{\partial w_j}{\partial x_k} \frac{\partial w_i}{\partial x_h} dx, \quad K = \text{const}. \tag{67}$$

Estimates (63), (64) follow from (65) - (67). The lemma is proved. □

Let us construct the asymptotic expansion in powers of ε for the solution of (56) with $\Phi \equiv 0$ in the form

$$\tilde{u}_\varepsilon(x) \cong \sum_{\ell=0}^{\infty} \varepsilon^{\ell+2} \sum_{\langle\alpha\rangle=\ell} N_\alpha^0(\xi)\, D^\alpha v(x), \quad \xi = \varepsilon^{-1} x, \tag{68}$$

where matrices N_α^0 and vector valued functions v are to be determined. Applying formally operator L_ε to $\tilde{u}_\varepsilon(x)$ we get

$$L_\varepsilon \tilde{u}_\varepsilon(x) \cong \sum_{\ell=0}^{\infty} \varepsilon^{\ell} \sum_{\langle\alpha\rangle=\ell} H_\alpha(\xi) D^\alpha v(x) \cong f(x), \tag{69}$$

where

$$H_0(\xi) \equiv \frac{\partial}{\partial \xi_k} (C^{kj}(\xi) \frac{\partial}{\partial \xi_j} N_0^0(\xi)),$$

$$H_{\alpha_1}(\xi) \equiv \frac{\partial}{\partial \xi_k} (C^{kj}(\xi) \frac{\partial}{\partial \xi_j} N_{\alpha_1}^0(\xi)) + \frac{\partial}{\partial \xi_k} (C^{k\alpha_1}(\xi) N_0^0(\xi)) + C^{\alpha_1 j}(\xi) \frac{\partial N_0^0(\xi)}{\partial \xi_j};$$

$$H_{\alpha_1 \ldots \alpha_s}(\xi) \equiv \frac{\partial}{\partial \xi_k} (C^{kj}(\xi) \frac{\partial}{\partial \xi_j} N_{\alpha_1 \ldots \alpha_s}^0(\xi)) + \frac{\partial}{\partial \xi_k} (C^{k\alpha_1}(\xi) N_{\alpha_2 \ldots \alpha_s}^0(\xi)) +$$

$$+ C^{\alpha_1 j}(\xi) \frac{\partial}{\partial \xi_j} N_{\alpha_2 \ldots \alpha_s}^0(\xi) + C^{\alpha_1 \alpha_2}(\xi) N_{\alpha_3 \ldots \alpha_s}^0, \quad s \geq 2.$$

Similarly to sections 1 and 2 we seek N_α^0 in the form $N_\alpha^0 = N_\alpha^1 + N_\alpha^2$, where N_α^1 are matrices whose elements are 1-periodic in ξ functions from $\hat{H}_0^1(Q \setminus G^0)$ and matrices $N_\alpha^2(\varepsilon^{-1}x)$ have elements decaying exponentially with the increase of the distance from $\partial\Omega$.

We introduce the following notation

$$T_0^1 = I; \; T_0^2 = 0; \; T_{\alpha_1}^q(\xi) = -\frac{\partial}{\partial \xi_k} (C^{k\alpha_1}(\xi) N_0^q(\xi)) - C^{\alpha_1 j}(\xi) \frac{\partial N_0^q(\xi)}{\partial \xi_j},$$

$$q = 1,2 \; ;$$

$$T_{\alpha_1 \ldots \alpha_1}^q(\xi) = -\frac{\partial}{\partial \xi_k} (C^{k\alpha_1}(\xi) N_{\alpha_2 \ldots \alpha_\ell}^q(\xi)) - C^{\alpha_1 j}(\xi) \frac{\partial}{\partial \xi_j} N_{\alpha_2 \ldots \alpha_\ell}^q(\xi) -$$

$$- C^{\alpha_1 \alpha_2}(\xi) N_{\alpha_3 \ldots \alpha_\ell}^q(\xi), \; \ell \geq 2, \; q = 1,2.$$

Let us define matrices $N_\alpha^1(\xi)$ as weak 1-periodic in ξ solutions of the problems

$$\frac{\partial}{\partial \xi_k} (C^{kj}(\xi) \frac{\partial N_\alpha^1(\xi)}{\partial \xi_j}) = T_\alpha^1(\xi) \quad \text{in } \mathbb{R}^n \setminus G_1,$$

$$N_\alpha^1(\xi) = 0 \text{ on } \partial G_1, \tag{70}$$

and matrices $N_\alpha^2(\xi)$ as weak solutions of the problems

$$\frac{\partial}{\partial \xi_k}\left(C^{kj}(\xi)\frac{\partial N_\alpha^2}{\partial \xi_j}\right) = T_\alpha^2(\xi) \text{ in } \varepsilon^{-1}\Omega_\varepsilon,$$

$$N_\alpha^2(\xi) = -N_\alpha^1(\xi) \text{ on } \partial(\varepsilon^{-1}\Omega_\varepsilon) \tag{71}$$

It is easy to prove by induction that N_α^1 and N_α^2 exist. In order to show that $N_\alpha^2(\varepsilon^{-1}x)$ are of boundary layer type in Ω_ε we shall need the following

THEOREM 5. Let $\tau(x)$ and Ω' be the same as in theorem 1 and $U(x)$ be a weak solution of problem (62) where f^0, $f^j \in L_2(\Omega_\varepsilon)$, $j = 1,\dots,n$, $\Phi \in H^1(\Omega_\varepsilon)$, $\Phi = 0$ on $\overline{\Omega'} \cap \partial\Omega_\varepsilon$. Then

$$\varepsilon^{-2}\int_{\Omega_\varepsilon\cap\Omega'} |U|^2 \exp\left(\frac{\delta\tau}{\varepsilon}\right)dx + \int_{\Omega_\varepsilon\cap\Omega'} |\nabla U|^2 \exp\left(\frac{\delta\tau}{\varepsilon}\right)dx + \int_{\Omega_\varepsilon} E(U)\exp\left(\frac{\delta\tau}{\varepsilon}\right)dx \leqslant$$

$$\leqslant K\left(\int_{\Omega_\varepsilon} E(U)\,dx + \varepsilon^2\int_{\Omega_\varepsilon\cap\Omega'} |f^0|^2\exp\left(\frac{\delta\tau}{\varepsilon}\right)dx + \int_{\Omega_\varepsilon\cap\Omega'} (f^j,f^j)\exp\left(\frac{\delta\tau}{\varepsilon}\right)dx\right), \tag{72}$$

where constants K, δ are positive and do not depend on ε, U;

$E(U) \equiv \sum_{i,j=1}^{n} \left(\frac{\partial U_j}{\partial x_i} + \frac{\partial U_i}{\partial x_j}\right)^2$, $(f,g) = f_i g_i$.

Proof. Let us take $v = (e^{\mu\tau}-1)U$ in the integral identity for $U(x)$ where μ is a parameter to be chosen later. We then have

$$\int_{\Omega_\varepsilon} c_{ij}^{hk}\frac{\partial U_i}{\partial x_h}\frac{\partial U_j}{\partial x_k}e^{\mu\tau}dx = -\int_{\Omega_\varepsilon} c_{ij}^{hk}\frac{\partial U_j}{\partial x_k}\mu\frac{\partial \tau}{\partial x_h}U_i e^{\mu\tau}dx - \int_{\Omega_\varepsilon}(f^0,U)(e^{\mu\tau}-1)dx +$$

$$+\int_{\Omega_\varepsilon}\left(f^j,\frac{\partial U}{\partial x_j}\right)(e^{\mu\tau}-1)dx - \int_{\Omega_\varepsilon}\mu\frac{\partial\tau}{\partial x_j}(f^j,U)e^{\mu\tau}dx + \int_{\Omega_\varepsilon} c_{ij}^{hk}\frac{\partial U_j}{\partial x_k}\frac{\partial U_i}{\partial x_h}dx .$$

Taking into account conditions (57) and the fact that $\tau = 0$ outside of Ω' we get

$$\int_{\Omega_\varepsilon} E(U)e^{\mu\tau}dx \leqslant c_1\mu\Big(\int_{\Omega_\varepsilon\cap\Omega'} E(U)e^{\mu\tau}dx\Big)^{1/2}\Big(\int_{\Omega_\varepsilon\cap\Omega'} |U|^2e^{\mu\tau}dx\Big)^{1/2} +$$

$$+\Big(\int_{\Omega_\varepsilon\cap\Omega'} |f^0|^2e^{\mu\tau}dx\Big)^{1/2}\Big(\int_{\Omega_\varepsilon\cap\Omega'} |U|^2e^{\mu\tau}dx\Big)^{1/2} +$$

$$+ c_2\Big(\int_{\Omega_\varepsilon\cap\Omega'} (f^j,f^j)e^{\mu\tau}dx\Big)^{1/2}\Big(\int_{\Omega_\varepsilon\cap\Omega'} |\nabla U|^2e^{\mu\tau}dx\Big)^{1/2} +$$

$$+\mu c_3\Big(\int_{\Omega_\varepsilon\cap\Omega'} (f^j,f^j)e^{\mu\tau}dx\Big)^{1/2}\Big(\int_{\Omega_\varepsilon\cap\Omega'} |U|^2e^{\mu\tau}dx\Big)^{1/2} + c_4\int_{\Omega_\varepsilon} E(U)dx.$$

Using the inequality obtained from (58) by the transformation $\xi = \varepsilon^{-1}x$ we find for every set $\omega_z = \varepsilon(Q\setminus G^0) + \varepsilon z \subset \Omega_\varepsilon\cap\Omega'$ that

$$\int_{\omega_z}|U|^2e^{\mu\tau}dx \leqslant c_5\varepsilon^2\int_{\omega_z}\sum_{i,j=1}^{n}\left(\frac{\partial(U_ie^{\mu\tau/2})}{\partial x_j} + \frac{\partial(U_je^{\mu\tau/2})}{\partial x_i}\right)dx \leqslant$$

$$\leqslant c_6\varepsilon^2\int_{\omega_z} E(U)e^{\mu\tau}dx + c_6\varepsilon^2\mu^2\int_{\omega_z}|U|^2e^{\mu\tau}dx. \tag{74}$$

Taking $\mu = \sigma/2\sqrt{c_6}\varepsilon$, where $\sigma = \text{const}\in(0,1)$ and will be chosen later, we deduce from (74) that

$$\int_{\omega_z}|U|^2e^{\mu\tau}dx \leqslant c_7\varepsilon^2\int_{\omega_z} E(U)e^{\mu\tau}dx \quad \text{for any } \omega_z\subset\Omega_\varepsilon\cap\Omega',$$

and therefore

$$\int_{\Omega_\varepsilon\cap\Omega'}|U|^2e^{\mu\tau}dx \leqslant c_7\varepsilon^2\int_{\Omega_\varepsilon\cap\Omega'} E(U)e^{\mu\tau}dx. \tag{75}$$

Due to (58) we also have the inequality

$$\int_{\Omega_\varepsilon\cap\Omega'}|\nabla(e^{\frac{\mu\tau}{2}}U)|^2dx \leqslant M_1\int_{\Omega_\varepsilon\cap\Omega'} E(e^{\frac{\mu\tau}{2}}U)dx.$$

Hence

$$\int_{\Omega_\varepsilon\cap\Omega'} |\nabla U|^2 e^{\mu\tau} dx \leq c_8 \Big[\mu \Big(\int_{\Omega_\varepsilon\cap\Omega'} |U|^2 e^{\mu\tau} dx \Big)^{1/2} \Big(\int_{\Omega_\varepsilon\cap\Omega'} |\nabla U|^2 e^{\mu\tau} dx \Big)^{1/2} +$$

$$+ \mu^2 \int_{\Omega_\varepsilon\cap\Omega'} |U|^2 e^{\mu\tau} dx + \int_{\Omega_\varepsilon\cap\Omega'} E(U) e^{\mu\tau} dx \Big].$$

Since $\mu = \sigma/2\sqrt{c_6}\varepsilon$, this inequality together with (75) yields

$$\int_{\Omega_\varepsilon\cap\Omega'} |\nabla U|^2 e^{\mu\tau} dx \leq c_9(\sigma+\sigma^2) \int_{\Omega_\varepsilon\cap\Omega'} |\nabla U|^2 e^{\mu\tau} dx + c_9 \int_{\Omega_\varepsilon\cap\Omega'} E(U) e^{\mu\tau} dx.$$

Therefore for all $\sigma \leq \min(1, 1/2c_9)$ we obtain the inequality

$$\int_{\Omega_\varepsilon\cap\Omega'} |\nabla U|^2 e^{\mu\tau} dx \leq c_{10} \int_{\Omega_\varepsilon\cap\Omega'} E(U) e^{\mu\tau} dx. \tag{76}$$

It follows from (73), (75), (76) that

$$\int_{\Omega_\varepsilon} E(U) e^{\mu\tau} dx \leq c_{11} \Big[\mu\varepsilon \int_{\Omega_\varepsilon\cap\Omega'} E(U) e^{\mu\tau} dx +$$

$$+ \varepsilon \Big(\int_{\Omega_\varepsilon\cap\Omega'} |f^0|^2 e^{\mu\tau} dx \Big)^{1/2} \Big(\int_{\Omega_\varepsilon\cap\Omega'} E(U) e^{\mu\tau} dx \Big)^{1/2} +$$

$$+ \Big(\int_{\Omega_\varepsilon\cap\Omega'} (f^j, f^j) e^{\mu\tau} dx \Big)^{1/2} \Big(\int_{\Omega_\varepsilon\cap\Omega'} E(U) e^{\mu\tau} dx \Big)^{1/2} +$$

$$+ \mu\varepsilon \Big(\int_{\Omega_\varepsilon\cap\Omega'} (f^j, f^j) e^{\mu\tau} dx \Big)^{1/2} \Big(\int_{\Omega_\varepsilon\cap\Omega'} E(U) e^{\mu\tau} dx \Big)^{1/2} + \int_{\Omega_\varepsilon} E(U) dx \Big] \tag{77}$$

Chosing σ sufficiently small (but independent of ε) and taking into account that $\mu = \sigma/2\sqrt{c_6}\varepsilon$ we get from (75) - (77) the estimate (72). The theorem is proved.

<u>LEMMA 8</u>. Matrices $N_\alpha^j(\varepsilon^{-1}x)$ satisfy the following inequalities

$$\varepsilon^2 \int_{\Omega_\varepsilon} |\nabla_x N_\alpha^j|^2 dx + \int_{\Omega_\varepsilon} |N_\alpha^j|^2 dx \leqslant c_\alpha \ , \ \langle\alpha\rangle \geqslant 0, \tag{78}$$

where constants c_α do not depend on ε. □

The proof of this lemma is based on lemma 7 and is similar to the proof of lemma 2.

THEOREM 6. Let $u_\varepsilon(x)$ be a weak solution of problem (56) and $\Phi = 0$ on $\partial\Omega_\varepsilon$, $f \in C^{s+2}(\bar{\Omega})$. Let

$$u_\varepsilon^s(x) = \sum_{\ell=0}^{s} \varepsilon^{\ell+2} \sum_{\langle\alpha\rangle=\ell} N_\alpha^0(\varepsilon^{-1}x) D^\alpha f(x),$$

$$v_\varepsilon^s(x) = \sum_{\ell=0}^{s} \varepsilon^{\ell+2} \sum_{\langle\alpha\rangle=\ell} N_\alpha^1(\varepsilon^{-1}x) D^\alpha f(x),$$

where $N_\alpha^0 = N_\alpha^1 + N_\alpha^2$, N_α^1, N_α^2 are weak solutions of problems (70), (71) respectively. Then

$$\| u_\varepsilon(x) - u_\varepsilon^s(x) \|_{1,\Omega_\varepsilon} \leqslant c\varepsilon^{s+1}, \tag{79}$$

$$\| u_\varepsilon(x) - v_\varepsilon^s(x) \|_{1,\Omega_\varepsilon \cap \Omega^0} \leqslant c_1 \varepsilon^{s+1}, \tag{80}$$

where constants c, c_1 do not depend on ε; Ω^0 is a subdomain of Ω, such that $\bar{\Omega}^0 \subset \Omega$.

Let Ω^1 be a subdomain of Ω and $u_\varepsilon(x)$ be a solution of problem (56) with $f \equiv 0$ in Ω, $\Phi \equiv \Phi_\varepsilon = 0$ on $\Omega^1 \cap \partial\Omega_\varepsilon$. Then for any subdomain $\Omega^2, \overline{\Omega^2} \subset \Omega^1$, there exist positive constants c_2, δ independent of ε and such that

$$\| u_\varepsilon(x) \|_{1,\Omega_\varepsilon \cap \Omega^2} \leqslant c_2 \| \Phi_\varepsilon \|_{1,\Omega_\varepsilon} \exp(-\frac{\delta}{\varepsilon}), \tag{81}$$

Proof. Estimate (81) is a consequence of theorem 5 and lemma 7. Let us prove estimates (79), (80). As has been done in sections 1 and 2, one can easily verify that $W \equiv u_\varepsilon - u_\varepsilon^s$ is a weak solution of the problem

$$L_\varepsilon(W) = \varepsilon^{s+1} \sum_{\alpha_1,\dots,\alpha_{s+1}=1}^{n} C^{\alpha_1 k} \frac{\partial}{\partial \xi_k} N^0_{\alpha_2\dots\alpha_{s+1}} D^{\alpha_1\dots\alpha_{s+1}} f +$$

$$+ \varepsilon^{s+1} \frac{\partial}{\partial \xi_h} (C^{h\alpha_1} N^0_{\alpha_2\dots\alpha_{s+1}}) D^{\alpha_1\dots\alpha_{s+1}} f +$$

$$+ \varepsilon^{s+1} \sum_{\alpha_1,\dots,\alpha_{s+1}=1}^{n} C^{\alpha_1\alpha_2} N^0_{\alpha_3\dots\alpha_{s+1}} D^{\alpha_1\dots\alpha_{s+1}} f +$$

$$+ \varepsilon^{s+2} \sum_{\alpha_1,\dots,\alpha_{s+2}=1}^{n} C^{\alpha_1\alpha_2} N^0_{\alpha_3\dots\alpha_{s+2}} D^{\alpha_1\dots\alpha_{s+2}} f ,$$

$$W|_{\partial\Omega_\varepsilon} = 0 \tag{82}$$

Using lemma 7 for W and lemma 8 to estimate the right-hand side of (82) we get inequalities (79). In the same way as in sections 1 and 2 on the basis of theorem 5 we obtain the following estimates for $N^2_\alpha(\varepsilon^{-1}x)$:

$$\|N^2_\alpha(\varepsilon^{-1}x)\|_{1,\Omega_\varepsilon\cap\Omega^0} \leqslant c_\alpha \exp(-\gamma/\varepsilon), \tag{83}$$

where constants c_α, γ are positive and do not depend on ε. Estimates (83) and (79) imply (80). The theorem is proved.

4. Some Generalizations

Analysing the proof of theorems 1, 3, 5 one can easily see that estimates similar to (17), (44), (72) can also be obtained in the case of some non-periodic structures. Suppose that a subdomain $\Omega' \subset \Omega$ is such that

$\bar{\Omega}' = \bigcup_{s=1}^{d_\varepsilon} \overline{B^\varepsilon_s}$ where B^ε_s are bounded domains in $\mathbb{R}^n$ such that $B^\varepsilon_i \cap B^\varepsilon_j = \emptyset$ for $i \neq j$. Suppose also that Γ^ε_s, $s = 1,\dots,d_\varepsilon$, are closed sets, $\Gamma^\varepsilon_s \subset \overline{B^\varepsilon_s}$ and for each $v \in C^1(\overline{B^\varepsilon_s})$, $v = 0$ in a neighbourhood of Γ^ε_s the Friedrichs inequality

$$\int_{B^\varepsilon_s} |v|^2 dx \leqslant c^* \varepsilon^2 \int_{B^\varepsilon_s} |\nabla v|^2 dx$$

holds with a constant c^* independent of s, v, ε.

Let $\tau(x)$ be a function in $C^2(\bar{\Omega})$, $\tau = 0$ in $\Omega\setminus\Omega'$, $\tau \geq 0$ in Ω.

THEOREM 7. Let $U(x) \in H^1(\Omega_\varepsilon)$ be a weak solution of the boundary value problem

$$\frac{\partial}{\partial x_j}\left(a_{ij}(x,\varepsilon)\frac{\partial U}{\partial x_i}\right) = f \quad \text{in } \Omega_\varepsilon = \Omega\setminus\bigcup_{s=1}^{d_\varepsilon}\Gamma_s^\varepsilon\,, \quad U = \Phi \quad \text{on } \partial\Omega_\varepsilon$$

where $\Phi \in H^1(\Omega_\varepsilon)$, $\Phi = 0$ on $\Omega' \cap \partial\Omega_\varepsilon$, $f \in L_2(\Omega_\varepsilon)$; $a_{ij}(x,\varepsilon)$ satisfy the conditions

$$\lambda_1|\xi|^2 \leq a_{ij}(x,\varepsilon)\xi_i\xi_j \leq \lambda_2|\xi|^2,$$

with positive constants λ_1, λ_2 independent of ε. Then for $U(x)$ estimate (17) is valid with constants $K>0$, $\delta>0$ depending only on c^*, λ_1, λ_2, $\|\tau\|_{C^2(\bar{\Omega})}$.

If $f \equiv 0$ in $\Omega_\varepsilon \cap \Omega'$, $\Phi = 0$ on $\bar{\Omega}' \cap \partial\Omega_\varepsilon$, then for any subdomain $\Omega^0 \subset \Omega'$ such that $\bar{\Omega}^0 \subset \Omega'$ the solution $U(x)$ satisfies the inequality

$$\|U\|_{1,\Omega^0\cap\Omega_\varepsilon} \leq C\left[\,\|\Phi\|_{1,\Omega_\varepsilon} + \|f\|_{L_2(\Omega_\varepsilon\setminus\Omega')}\right]\exp\left(-\frac{\delta}{\varepsilon}\right),$$

where constants $C>0$, $\delta>0$ depend only on c^*, λ_1, λ_2, Ω^0. □

This theorem is proved in the same way as theorem 1, if one replaces the sets ω_z by $B_s^\varepsilon \setminus \Gamma_s^\varepsilon$.

Similarly for the biharmonic equation we have

THEOREM 8. Let $U(x) \in H^2(\Omega_\varepsilon)$ be a weak solution of the problem

$$\Delta^2 U = f_0 + \frac{\partial f_i}{\partial x_i} \quad \text{in } \Omega_\varepsilon = \Omega\setminus\bigcup_{s=1}^{d_\varepsilon}\Gamma_s^\varepsilon\,,$$

$$U = \Phi, \quad \frac{\partial U}{\partial \nu} = \frac{\partial \Phi}{\partial \nu} \quad \text{on } \partial\Omega_\varepsilon\,,$$

where $\Phi \in H^2(\Omega_\varepsilon)$, $\Phi = \frac{\partial\Phi}{\partial\nu} = 0$ on $\Omega' \cap \partial\Omega_\varepsilon$. Then for $U(x)$ inequality (44) holds with constants $K_0 > 0$, $\delta > 0$ depending on c^*, $\|\tau\|_{C^2(\bar{\Omega})}$ only.

If $f_0 \equiv f_i \equiv 0$, $i = 1,\ldots,n$, in $\Omega_\varepsilon \cap \Omega'$, $\Phi = \frac{\partial\Phi}{\partial\nu} = 0$ on $\Omega' \cap \partial\Omega_\varepsilon$ then for any subdomain $\Omega^0 \subset \Omega'$ such that $\overline{\Omega^0} \subset \Omega'$ the solution $U(x)$ satisfies the estimate

$$\|U\|_{2,\Omega^0\cap\Omega_\varepsilon} \leq c\left[\|\Phi\|_{2,\Omega} + \sum_{j=0}^{n} \|f_j\|_{L_2(\Omega_\varepsilon\setminus\Omega')}\right] \exp\left(-\frac{\delta}{\varepsilon}\right),$$

where $c > 0$, $\delta > 0$ are constants depending only on c^*, Ω^0. □

Consider now the case of the system of linear elasticity. Suppose that the sets Γ_s^ε, $s = 1,\ldots,d_\varepsilon$, are such that for every vector valued function $v \in C^1(\bar{B}_s^\varepsilon)$, $v = 0$ in a neighbourhood of Γ_s^ε, the following inequalities are satisfied

$$\int_{B_s^\varepsilon} |v|^2 dx \leq c_1 \varepsilon^2 \int_{B_s^\varepsilon} E(v)\,dx, \quad \int_{B_s^\varepsilon} |\nabla v|^2 dx \leq c_2 \int_{B_s^\varepsilon} E(v)\,dx,$$

where constants c_1, c_2 do not depend on s and ε.

<u>THEOREM 9</u>. Let $U(x)$ be a weak solution of the boundary value problem for the elasticity system

$$\frac{\partial}{\partial x_h}\left(A^{hk}(x,\varepsilon)\frac{\partial U}{\partial x_k}\right) = f^0 + \frac{\partial f^i}{\partial x_i} \quad \text{in } \Omega_\varepsilon = \Omega \setminus \bigcup_{s=1}^{d_\varepsilon} \Gamma_s^\varepsilon,$$

$$U = \Phi \text{ on } \partial\Omega_\varepsilon,$$

where $f^0, f^i \in L_2(\Omega_\varepsilon)$, $i = 1,\ldots,n$; $\Phi \in H^1(\Omega)$, $\Phi = 0$ on $\Omega' \cap \partial\Omega_\varepsilon$, the elements $a_{ij}^{hk}(x,\varepsilon)$ of matrices $A^{hk}(x,\varepsilon)$ are uniformly bounded in ε by a constant M and satisfy conditions similar to (57) with constants $\tilde{A}_1$, $\tilde{A}_2$ independent of ε. Then for $U(x)$ estimate (72) holds with constants $K > 0$, $\delta > 0$ depending only on c_1, c_2, $\tilde{A}_1$, $\tilde{A}_2$, $M, \|\tau\|_{C^1(\bar{\Omega})}$.

If $f^0 \equiv f^i \equiv 0$, $i = 1,\ldots,n$, in $\Omega_\varepsilon \cap \Omega'$, $\Phi = 0$ on $\Omega' \cap \partial\Omega_\varepsilon$ then for any subdomain $\Omega^0 \subset \Omega'$ such that $\overline{\Omega^0} \subset \Omega'$ the solution U(x) satisfies the inequality

$$\|U\|_{1,\Omega^0\cap\Omega_\varepsilon} \leqslant C\Big[\|\Phi\|_{1,\Omega_\varepsilon} + \sum_{j=0}^{n} \|f^j\|_{L_2(\Omega_\varepsilon\setminus\Omega')}\Big]\exp(-\frac{\delta}{\varepsilon}) ,$$

where constants $C>0$, $\delta>0$ depend only on c_1, c_2, $\tilde{A}_1$, $\tilde{A}_2$, M, Ω^0.

References

[1] J.L. Lions, Some methods in the mathematical analysis of systems and their control, Science Press, Beijing, China, Gordon and Breach, Inc. New York (1981).

[2] J.L. Lions, Asymptotic expansions in perforated media with a periodic structures, The Rocky Mountain Journ. of Math.,(1980), v. 10, no. 1, p. 125-140.

[3] V.A. Marchenko, E.Ya. Hruslov, Boundary value problems in domains with a finely granulated boundary. Kiev, Naukova dumka, (1974).

[4] O.A. Oleinik, G.P. Panasenko, G.A. Yosifian, Homogenization and asymptotic expansions for solutions of the elasticity system with rapidly oscillating periodic coefficients, Applicable Analysis (1983), v.15, no. 1-4, p. 15-32.

[5] O.A. Oleinik, G.P. Panasenko, G.A. Yosifian, Asymptotic expansion for solutions of the elasticity system in perforated domains. Matem. Sbornok, (1983), v. 120, no. 1, p. 22-41.

[6] O.A. Oleinik, On homogenization problems. Trends and Applications of Pure Mathematics to Mechanics, Lecture Notes in Physics, no. 195, Springer Verlag (1984), p. 248-272.

[7] O.A. Oleinik, A.S.Shamaev, G.A. Yosifian, Homogenization of eigenvalues and eigenfunctions of the boundary value problem of elasticity in a perforated domain, Vestnik Mosc. Univ., ser. 1, Mat., Mech., (1983), no. 4, p. 53-63.

[8] O.A. Oleinik, A.S. Shamaev, G.A. Yosifian, On the convergence of the energy, stress tensors and eigenvalues in homogenization problems arising in elasticity, Dokladi AN SSSR (1984), v. 274, no. 6, p. 1329-1333.

[9] O.A. Oleinik, G.A. Yosifian, On homogenization of the elasticity system with rapidly oscillating coefficients in perforated domains. In the book "N.E. Kochin and the development of Mechanics", M. Nauka, (1984), p. 237-249.

[10] O.A. Oleinik, G.A. Yosifian, On the asymptotic behaviour at infinity of solutions in linear elasticity, Archive Rat. Mech. and Analysis (1982), v. 78, no. 1, p. 29-53.

[11] O.A. Oleinik, G.A. Yosifian, On the behaviour at infinity of solutions of second order elliptic equations in domains with non-compact boundaries, Matem. Sbornik (1980) 112 (154), no. 4, p. 588-610.

[12] V.A. Kondratiev, On the solvability of the first boundary value problem for strongly elliptic equations. Trans. of the Moscow Math. Soc. (1967), v. 16, p. 293-318.

[13] V.G. Maz'ya, Polyharmonic capacity in the theory of the first boundary value problem. Sibirsk. Mat. Journ. (1965), v. 6., no. 1, p. 127-148.

[14] P.P. Mosolov, V.P. Miasnikov, A proof of Korn's inequality, Dokladi AN SSSR, (1971), v. 201, no. 1, p. 36-39.

[15] G. Fichera, Existence theorems in elasticity. Handbuch der Phisik, Band VI/a2, Springer Verlag.

O.A. Oleinik , A.S. Shamaev & G.A. Yosifian
Moscow University
M.G.U. 78
"K" Kvartal 133
Moscow 117234
Soviet Union

R TAHRAOUI

Quelques remarques sur le contrôle des valeurs propres

1. Introduction

Il existe de nombreux travaux traitant du contrôle des fonctions et valeurs propres pour les opérateurs elliptiques. Les premiers résultats semblent remonter à 1955 avec le travail de M.G. Krein [K] qui, motivé par l'étude des zones de stabilité de l'équation $y''+\lambda p(x)y = 0$ - $(-\infty < x < +\infty)$- où p est une fonction périodique, s'était intéressé aux deux problèmes suivants :

$$\inf\{\lambda_n(v),\ v\in U\}\ ,\ \sup\ \{\lambda_n(v),\ v\in U\}$$

où $\lambda_n = \lambda_n(v)$ est la $n^{\text{ième}}$ valeur propre de

$$\begin{cases} -y'' = \lambda_r.v(x).y \\ y(0) = y(1) = 0 \end{cases}$$

avec U donné par

$$U = \{v\in L^\infty(0,1)/h\leqslant v\leqslant H,\quad \int_0^1 v(x)dx = M\},$$

h et H étant deux constantes positives. Les résultats obtenus sont précis : sup λ_n est donné explicitement en fonction des paramètres du problème et inf λ_n en fonction de ces mêmes paramètres et de la plus petite racine positive d'une équation transcendentale. Motivé par l'étude des vibrations d'une barre D.O. Banks [B] aborda les deux questions précédents pour l'opérateur d'ordre 4 suivant :

$$\begin{cases} y^{(4)}-\lambda.v(x).y = 0 \\ y(0) = y(1) = y''(0) = y''(1) = 0, \end{cases}$$

v appartenant à U défini ci-dessus. Nous pouvons remarquer que dans [K] comme dans [B] le problème dépend de manière convexe du contrôle; et dans ce cas la question de l'existence ne se pose pas. Dans un cadre plus général que dans [K,B], G. Zera [ZE] donne, outre plusieurs motivations mécaniques dans la recherche des formes optimales, des résultats d'existence en utilisant diverses techniques, notamment l'homogénéisation et les réarrangements, pour problèmes de la forme :

$$(P_1)\begin{cases} -(a(v)y')' = \lambda_1 . b(v) . y \\ \sup \lambda_1(v),\ \lambda_1 \text{ est la } 1^{\text{ière}} \text{ valeur propre,} \end{cases}$$

sous différentes conditions aux limites, le contrôle v appartenant à un ensemble convexe de type U. Signalons à titre d'exemple un résultat très significatif de [ZE] : (P_1) possède au moins une solution si la fonction $v \to a(v).b(v)$ est strictement monotone. Dans [J] C. Jouron étudie le problème

$$(P_2)\quad \begin{aligned} & -\text{div}\,(a(v).\nabla\omega) = \lambda_1 . b(v) . \omega \text{ dans } \Omega \subset R^n,\ n \geqslant 2 \\ & \omega/_{\partial\Omega} = 0 \\ & \sup\{\lambda_1(v),\ v \in U\} \end{aligned}$$

où b est convexe et a concave; le contrôle v appartient à un ensemble convexe U du même type que les précédents. Il donne un résultat d'existence et des conditions d'optimalité. Dans le même sens, on peut citer à titre d'exemples les travaux de F. Murat [MU_1], F. Mignot [MI_2]. On trouvera dans [J], [ZE] et [CH1] une bibliographie assez large sur le sujet. Si dans le problème (P_2) a n'est pas concave ou b n'est pas convexe le problème de l'existence reste posé; et il semble que des résultats d'existence du même type que celui énoncé ci-dessus pour (P_1) soient difficiles à obtenir. En effet, en dimension d'espace $n \geqslant 2$, d'une manière générale, si on se place dans une situation où la dépendance du contrôle est non convexe, l'existence d'une solution optimale n'est pas toujours assurée.
L. Tartar avait montré vers les années 1980-1981 [Tar1] que pour a(v) = v, b(v) = 1 et U l'ensemble non convexe suivant :

$$U := \{v \in L^\infty(\Omega) / v(x) \in \{\alpha,\beta\},\ \text{mesure}\ (v^{-1}(\alpha)) = \theta_1,$$

$$\text{mesure}\ (v^{-1}(\beta)) = \theta_2,\ \theta_i\ \text{donné tel que}\ \theta_1+\theta_2 = \text{mesure}\ (\Omega)\},$$

(P_2) n'admet pas, en général, de solution; il se produit un phénomène d'homogénéisation : il s'agit ici de contrôle de domaine. Pour une situation générale où le contrôle intervient dans les termes d'ordres deux de l'opérateur, on pourra se reporter à [MT] où des informations qualitatives sont données sur certaines solutions; on peut également consulter (G. Milton), [ALT1] pour des résultats qualitatifs. Dans ce travail nous nous intéressons au contrôle des valeurs propres pour les opérateur uniformément elliptiques d'ordre 2, avec une dépendance non convexe, et où le contrôle intervient au niveau des termes d'ordre zéro de l'opérateur différentiel considéré. Nous vous attacherons à dégager quelques situations où l'existence peut être démontrée; et nous donnerons quelques informations qualitatives sur la solution optimale. Nous aborderons également la question de l'unicité.

2. Notations et positions du problème

On se donne un ouvert Ω borné, régulier de $\mathbb{R}^n$ et trois fonctions α, β, a telles que

$$\begin{aligned} &\alpha,\beta \in L^\infty(\Omega), \\ &0 < \alpha_0 \leqslant \alpha(x) < \beta(x)\ \text{p.p.}\ x \in \Omega \end{aligned} \tag{2.1}$$

où α_0 est une constante,

$$\begin{aligned} &a : (x,t) \in \Omega \times \mathbb{R}^+ \to a(x,t) \in \mathbb{R}^+ \\ &a(x,t) \geqslant a_0 > 0 \end{aligned} \tag{2.2}$$

où a_o est une constante. Dans la suite ces fonctions seront supposées régulières. On définit l'ensemble des contrôles admissibles

$$U = U(\alpha,\beta,\gamma) = \{v \in L^\infty(\Omega) / \alpha \leqslant v \leqslant \beta,\ \int_\Omega v\,dx = \gamma > 0\} \tag{2.3}$$

qu'on suppose non vide i.e. que la constante γ vérifie

$$\int_\Omega \beta \, dx \geqslant \gamma \geqslant \int_\Omega \alpha \, dx.$$

L'état du système considéré, est défini par le couple (λ_1, w) solution, unique à une normalisation près, de :

$$\begin{cases} -\Delta\omega = \lambda_1 . a(x,v) . \omega \text{ dans } \Omega \\ \\ \omega/_{\partial\Omega} = 0, \; \omega > 0 \quad \text{ dans } \Omega, \end{cases} \tag{2.4}$$

où λ_1 désigne la première valeur propre de (2.4). Dans toute la suite nous noterons indifféremment - (si aucune ambiguîté n'est à craindre) -

$$\lambda_1 = \lambda_1(v) = \lambda_1(a,v) = \lambda_1(a) = \lambda_1(\mu) \tag{2.5}$$

la première valeur propre, où $\mu(x)$ est la fonction $a(x,v(x))$, et

$$\omega = \omega(v) = \omega(a) = \omega(a,v) \tag{2.6}$$

la fonction propre - (à une normalisation près) - correspondante.
La fonction coût est donnée par la fonctionelle

$$J(v) = \lambda_1(v).$$

Et le problème, dans sa généralité, s'enonce ainsi :

$$\begin{cases} \text{Trouver } u \text{ dans } U \text{ réalisant le suprémum de} \\ J(v) = \lambda_1(v) \\ \text{sur l'ensemble } U. \end{cases} \tag{P_a}$$

3. Existence de contrôle bang-bang

Pour la suite de l'étude, précisons ce que nous entendons par contrôle bang-bang.

DEFINITIONS 3.1. Un contrôle v appartenant à U est dit bang-bang si :

$$v(x) \in \{\alpha(x), \beta(x)\} \text{ p.p. } x \in \Omega .$$

Il s'agit, dans ce chapitre, de trouver des conditions suffisantes vérifiées par a, qui assurent l'existence d'un contrôle optimal, solution de (P_a), possédant la propriété bang-bang. Pour cela nous supposons les hypothèses suivantes :

$$\begin{aligned} &a(x,t) \geqslant L(x,t) = \theta(x).t + s(x) \geqslant L_0 > 0 \\ &(x,t) \in \Omega \times [\alpha(x), \beta(x)], \end{aligned} \tag{3.1}$$

et

$$\begin{aligned} &a(x,\alpha(x)) = L(x,\alpha(x)) \text{ p.p. } x \in \Omega, \\ &a(x,\beta(x)) = L(x,\beta(x)) \text{ p.p.} x \in \Omega, \end{aligned} \tag{3.2}$$

où $\theta = \dfrac{a(x,\beta)-a(x,\alpha)}{\beta-\alpha}$ satisfait

$$\begin{aligned} &\forall \lambda \in [\lambda_1(\nu_2), \lambda_1(\nu_1)] \\ &\frac{-\Delta(1/\sqrt{\theta(x)})}{\lambda.1/\sqrt{\theta(x)}} \notin [a(x,\alpha(x)), a(x,\beta(x))] \text{ p.p.} x \in \Omega, \end{aligned} \tag{3.3}$$

avec $\lambda_1(\nu_i)$ la première valeur propre de

$$\begin{cases} -\Delta\varphi = \lambda_1.\nu_i.\varphi \\ \varphi/_{\partial\Omega} = 0, \end{cases}$$

les fonctions ν_i étant définies par :

$$\begin{cases} \nu_1(x) = \inf\{a(x,t), t \in [\alpha(x),\beta(x)]\} & \text{p.p.} x \in \Omega, \\ \nu_2(x) = \sup\{a(x,t), t \in [\alpha(x),\beta(x)]\} & \text{p.p.} x \in \Omega. \end{cases}$$

DEFINITION 3.2. Une fonction $f(x)$ mesurable sur Ω, à valeurs réelles, est dite sans palier si pour tout nombre réel t on a :

$$|\{x \in \Omega / f(x) = t\}| = 0,$$

où $|E|$ désigne la mesure de Lebesque de tout mesurable E de Ω.

REMARQUE 3.1. L'hypothèse (3.3) entraine que pour tout v satisfaisant $\alpha \leqslant v \leqslant \beta$, la fonction propre $\omega = \omega(L,v)$ correspondante est telle que $\theta\omega^2$ n'a pas de palier. Elle est vérifiée si $\frac{1}{\sqrt{\theta}}$ est harmonique, et par exemple si θ est constante.

Dans le but de simplifier l'énoncé du théorème 3.1 ci-dessus et sa démonstration, nous supposerons que θ est une fonction positive vérifiant

$$|\{x \in \Omega / \theta(x) = 0\}| = 0.$$

Le cas où θ change de signe sera commenté en remarques 3.3.

THEOREME 3.1. Sous les hypothèses (2.1), (2.2), (3.1), (3.2) et (3.3), le problème (Pa) possède un contrôle optimal $\bar{u}$, ayant la propriété bang-bang. De façon plus précise, il existe un réel $t_o > 0$ et un mesurable $\Omega\alpha \subset \Omega$, défini à un ensemble de mesure nulle près, tels que

$$\{x \in \Omega / \theta(x)\bar{\omega}^2(x) > t_o\} \subseteq \Omega_\alpha \subseteq \{x \in \Omega / \theta(x)\bar{\omega}^2(x) \geqslant t_o\}, \tag{3.4}$$

$\bar{\omega}$ étant l'état optimal correspondant à $\bar{u}$ qui est caractérisé par :

$$\bar{u}(x) = \begin{cases} \alpha(x) & \text{p.p.} x \in \Omega_\alpha \\ \beta(x) & \text{p.p.} x \in \Omega_\beta = \Omega \setminus \Omega_\alpha . \end{cases} \tag{3.5}$$

De plus cette solution $(\bar{u}, \bar{\omega})$ est unique à une normalisation près.

Avant d'aborder la preuve du théorème ci-dessus, donnons quelques exemples.

EXEMPLE 1. Contrôle de domaine.

a, α, β sont supposés indépendants de x; alors il existe $t_o > 0$ tel que si l'on pose

$$D_o = \{x \in \Omega / \bar{\omega}(x) \geqslant t_o\},$$

le contrôle optimal $\bar{u}$, solution de (Pa) s'écrit

$$\bar{u}(x) = \begin{cases} \alpha \text{ sur } D_o \\ \beta \text{ sur } \Omega \setminus D_o; \end{cases}$$

et bien entendu on a :

$$\lambda_1(\bar{u}) = \lambda_1(D_o) = \sup\{\lambda_1(D)/D \text{ mesurable} \subset \Omega, \ |D| = \frac{\beta|\Omega|-\gamma}{\beta-\alpha}\},$$

où $\lambda_1(D)$ représente la première valeur propre de

$$\begin{cases} -\Delta\omega_D = \lambda_1(D).a(v_D).\omega_D \\ \omega_D/_{\partial\Omega} = 0, \ \omega_D > 0 \text{ dans } \Omega, \end{cases}$$

la fonction v_D étant définie par :

$$v_D(x) = \begin{cases} \alpha \text{ si } x \in D \\ \beta \text{ si } x \in \Omega \setminus D. \end{cases}$$

EXEMPLE 2. $n = 1$, $\Omega =]0,1[$.

α et β sont supposés constants et $a(x,t) = t$. Dans cette situation $\bar{u}$ est donnée explicitement :(cf [K],[B])

$$\bar{u}(x) = \begin{cases} \alpha \text{ si } x \in [r, 1-r] \\ \beta \text{ si } x \in [0, r[\ \cup \ ]1-r, 1] \end{cases}$$

$$r = \frac{\gamma-\alpha.1}{2(\beta-\alpha)}.$$

EXEMPLE 3. La situation radiale.

On suppose que Ω est le disque de centre 0 et de rayon $R>0$ -(pour simplifier on prend $n = 2$)- . Dans ce cas la solution optimale $(\bar{u},\bar{\omega})$ de (Pa) est radiale et le contrôle optimal $\bar{u}$ a pour expression :

$$\bar{u}(x) = \begin{cases} \alpha & \text{si } x \in D_o \\ \beta & \text{si } x \in \Omega \setminus D_o, \end{cases}$$

où D_o est le disque concentrique à Ω, de rayon R_o tel que

$$R_0^2 = \frac{\beta.\pi.R^2-\gamma}{(\beta-\alpha)\pi} .$$

Abordons maintenant la preuve du théorème.

Démonstration

i) l'existence :

Cette preuve repose sur la condition d'optimalité vérifié par la solution optimale. Puisque (3.1) a lieu, on a

$$\sup \{\lambda_1(L), v \in U\} \geqslant \sup \{\lambda_1(a),\ v \in U\},$$

et grâce à (3.2) on voit qu'il suffit de montrer que (P_L) admet un contrôle optimal $\bar{u}$ bang-bang. Pour cela on raisonne par l'absurde . Soit $(\bar{\omega},\bar{u})$ une solution optimale de (P_L); on montre que cette solution est caracterisée par la condition d'optimalité

$$\int_\Omega (v-\bar{u})\theta\bar{\omega}^{-2}dx \geqslant 0 \quad \forall v \in U. \tag{3.6}$$

Supposons que l'ensemble

$$\Omega' = \{x \in \Omega / \alpha(x) < \bar{u}(x) < \beta(x)\}$$

est de mesure non nulle i.e. que $\bar{u}$ n'est pas bang-bang.

Et posons

$$h(t) = \mu(\{x \in \Omega' / \theta(x)\bar{\omega}^{-2}(x) > t\})$$

où μ est la mesure positive définie par

$$\mu(E) = \int_E (\beta-\alpha)dx$$

pour tout mesurable $E \subset \Omega'$. Cette fonction h est continue puisque $\theta\bar{\omega}^{-2}$ est sans palier grâce à (3.3); ceci permet de montrer l'existence de deux mesurables E_1 et E_2 dans Ω', définis à un ensemble de mesure nulle près, satisfaisant :

$$\int_{E_1} \alpha \, dx + \int_{E_2} \beta \, dx = \int_{\Omega'} \bar{u} \, dx = \gamma' > 0,$$

$$\theta(x) \bar{\omega}^{-2}(x) < \theta(y)\bar{\omega}^{-2}(y) \quad \text{p.p. } (x,y) \in E_2 \times E_1$$

où E_1 est défini par l'existence d'un réel $t > 0$ tel que

$$E_1 = \{y \in \Omega' / \theta(y)\bar{\omega}^{-2}(y) > t\} \ ;$$

et

$$E_2 = \Omega' \setminus E_1, \ |E_i| > 0 \quad i = 1,2.$$

Enfin pour conclure, on montre, après quelques calculs, que le contrôle admissible

$$v_0(x) = \begin{cases} \alpha & \text{sur } E_1 \\ \beta & \text{sur } E_2 \\ \bar{u} & \text{sur } \Omega \setminus \Omega' \end{cases}$$

vérifie l'inégalité

$$\int_\Omega (v_0 - \bar{u})\theta\bar{\omega}^{-2} \, dx < 0$$

qui contredit (3.6). Ainsi $|\Omega'| = 0$, et tout contrôle optimal de (P_L) est bang-bang. Et en raisonnant encore par l'absurde on montre facilement qu'il existe $t_o > 0$ tel que

$$\bar{u}(x) = \begin{cases} \alpha(x) & \text{sur } \Omega_\alpha \\ \beta(x) & \text{sur } \Omega_\beta = \Omega \setminus \Omega_\alpha . \end{cases}$$

où

$$\Omega_\alpha = \{x \in \Omega / \theta(x)\bar{\omega}^2(x) > t_o\} \ .$$

Enfin la propriété bang-bang de $\bar{u}$ et (3.2)-(3.1) entrainent que $(\bar{\omega},\bar{u})$ est aussi solution optimale de (Pa).

ii) l'unicité :

La preuve de l'unicité utilise la propriété (3.4)-(3.5) du contrôle optimal. En effet d'après i) (Pa) admet une solution optimale (ω_1,u_1) avec u_1 bang-bang. Soit alors (ω_2,u_2) une autre solution de (Pa); on montre, en raisonnant par l'absurde et en utilisant l'égalité

$$\sup \{\lambda_1(a,v),\ v\in U\} = \sup \{\lambda_1(L,v),\ v\in U\},$$

que u_2 est bang-bang. Par conséquent (ω_2,u_2) est aussi solution optimale de (P_L); et le problème revient à montrer l'unicité de la solution de (P_L). Nous ferons la démonstration en raisonnant par l'absurde i.e. nous supposons $u_1 \not\equiv u_2$. Posons alors

$$\Omega_\alpha^i = \{x\in \Omega/\theta\omega_i^2(x) \geqslant t_i\},\ i = 1,2;$$

$$\Omega_\beta^i = \{x\in \Omega/\theta\omega_i^2(x) < t_i\},\ i = 1,2;$$

$$\Delta\ (\alpha,\beta) = \int_{\Omega_\alpha^1\setminus\Omega_\alpha^2} (\beta-\alpha)\ dx - \int_{\Omega_\alpha^2\setminus\Omega_\alpha^1} (\beta-\alpha)\ dx\ ;$$

nous distinguons deux cas

1) <u>$\Delta(\alpha,\beta) \geqslant 0$</u> :

Nous avons

$$\int_\Omega (L(x,u_2)-L(x,u_1))\omega_1^2\ dx = \int_{\Omega_\alpha^1\setminus\Omega_\alpha^2}(\beta-\alpha)\theta\omega_1^2 dx + \int_{\Omega_\alpha^2\setminus\Omega_\alpha^1}(\alpha-\beta)\theta\omega_1^2 dx; \quad (3.7)$$

comme $\Delta(\alpha,\beta)$ est positif et $u_1 \not\equiv u_2$, on peut affirmer que

$$|\Omega_\alpha^1 \setminus \Omega_\alpha^2| > 0.$$

Ceci permet de minorer strictement le second membre de (3.7) :

$$\int_\Omega (L(x,u_2)-L(x,u_1))\omega_1^2 dx > t_1\Delta(\alpha,\beta) \geqslant 0 \tag{3.8}$$

qui entraine

$$\lambda_1(L,u_1) > \lambda_1(L,u_2);$$

ce qui contredit l'optimalité de u_2.

2) <u>$\Delta(\alpha,\beta) \leqslant 0$</u> :

Ce cas se traite comme précédemment en permutant les rôles de u_1 et u_2; ce qui achève la démonstration du théorème. □

REMARQUE 3.2.

Moyennant une modification convenable de (3.3), le théorème (3.1) reste valable si l'on remplace l'ensemble des contrôles admissibles U donné en (2.3) par l'ensemble non convexe

$$U' = U'(\alpha,\beta,\gamma) = \{v \in L^\infty(\Omega)/\alpha \leqslant v \leqslant \beta,\ \int_\Omega f(v)\,dx = \gamma\} \tag{3.9}$$

où f est une fonction de $\mathbb{R}^+$ dans $\mathbb{R}^+$, convexe, strictement croissante. Ceci se voit en faisant un changement de fonction contrôle; i.e. on pose :

$$V = f(v)$$
$$\tilde{a}(x,V) = a(x, f^{-1}(V))$$
$$\tilde{U}(f(\alpha),f(\beta),\gamma) = \tilde{U} = \{V \in L^\infty(\Omega)/f(\alpha) \leqslant V \leqslant f(\beta),\ \int_\Omega V dx = \gamma\};$$

et le problème (Pa) peut se formuler comme suit :

$$\begin{cases} \text{Trouver } \bar{V} \in \tilde{U} \text{ tel que} \\ \lambda_1(\bar{V}) = \sup\{\lambda_1(V),\ V \in \tilde{U}\} \end{cases}$$

où $\lambda_1(V)$ est la première valeur propre de

$$\begin{cases} -\Delta\omega = \lambda_1(V).\tilde{a}(x,V).\omega \text{ dans } \Omega \\ \omega/_{\partial\Omega} = 0, \ \omega > 0 \qquad \text{dans } \Omega \end{cases}$$

Si l'on note

$$\tilde{\theta}(x) = \frac{a(x,\beta(x)) - a(x,\alpha(x))}{f(\beta(x)) - f(\alpha(x))}, \ x \in \Omega,$$

$$\tilde{s}(x) = \theta(x).\frac{-\beta f(\alpha) + \alpha f(\beta)}{f(\beta) - f(\alpha)} + s(x),$$

on a l'hypothèse (3.1) - (3.2) réalisée pour $\tilde{U}$:

$$\tilde{a}(x,T) \geqslant \tilde{\theta}(x).T + \tilde{s}(x) = \tilde{L}(x,T)$$

$$\tilde{a}(x,f(\alpha)) = \tilde{L}(x,f(\alpha))$$

$$\tilde{a}(x,f(\beta)) = \tilde{L}(x,f(\beta)).$$

Et il suffit d'imposer la condition (3.3) pour $\tilde{\theta}$ pour que toutes les conditions d'application du théorème (3.1) soient réunies. Signalons enfin que nous obtenons un exemple intéressant pour α, β constants, $a(x,t) = a(t)$ satisfaisant (3.1) - (3.2) et $f(t) = t^p$ avec $p > 1$. Et il est a remarquer que dans ce cas (3.3) a lieu automatiquement.

REMARQUES 3.3.

i) Nous considérons ici la situation où θ change de signe; nous posons

$$\Omega+ = \{x \in \Omega / \theta(x) > 0\},$$

$$\Omega- = \{x \in \Omega / \theta(x) < 0\},$$

$$\Omega_0 = \{x \in \Omega / \theta(x) = 0\}.$$

Alors pour tout contrôle optimal $\bar{u}$ de (P_L), il existe $t_1 > 0$ et $t_2 > 0$ tels que si l'on définit

$$\Omega_+^\alpha = \{x \in \Omega_+ / \theta\bar{\omega}^2 > t_1\},$$

$$\Omega_+^\beta = \{x \in \Omega_+ / \theta\bar{\omega}^2 \leqslant t_1\},$$

$$\Omega_-^\alpha = \{x \in \Omega_- / -\theta\bar{\omega}^2 < t_2\},$$

$$\Omega_-^\beta = \{x \in \Omega_- / -\theta\bar{\omega}^2 \geqslant t_2\},$$

le contrôle optimal s'écrit

$$\bar{u} = \begin{cases} \alpha \text{ sur } \Omega_+^\alpha \\ \beta \text{ sur } \Omega_+^\beta \\ \alpha \text{ sur } \Omega_-^\alpha \\ \beta \text{ sur } \Omega_-^\beta \quad ; \end{cases}$$

et $\bar{u}$ est indéterminé sur Ω_0 si $|\Omega_0| > 0$; en effet tout contrôle admissible v_0 vérifiant les deux conditions

$$v_0 = \bar{u} \quad \text{sur } \Omega \setminus \Omega_0,$$

et

$$\int_{\Omega_0} u \, dx = \int_{\Omega_0} v_0 \, dx,$$

est optimal.

ii) Plaçons-nous maintenant dans le cas où l'hypothèse (3.3) fait défaut; alors tout contrôle optimal $\bar{u}$ de (P_L) satisfaisant $|\Omega'| > 0$ avec

$$\Omega' = \{x \in \Omega / \alpha(x) < \bar{u}(x) < \beta(x)\},$$

vérifie

$$\theta(x)\bar{\omega}^2(x) = C_o \text{ p.p. } x \in \Omega';$$

la constante C_o est la même pour toutes les composantes connexes de Ω'. Le contrôle est caractérisé par

$$\bar{u} = \begin{cases} \alpha \text{ si } \theta\bar{\omega}^{-2} > C_o \\ \beta \text{ si } \theta\bar{\omega}^{-2} < C_o , \end{cases}$$

dans $\Omega \setminus \Omega'$ et dans la zone $\Omega' = \{x \in \Omega / \theta(x)\bar{\omega}^{-2}(x) = C_o\}$ nous ne possédons aucune information sur $\bar{u}$. Dans [MT] F. Murat et L. Tartar font une remarque similaire pour le problème du contrôle de la rigidité à la torsion; et signalons que [MT] contient également des informations qualitatives sur d'autres problèmes de contrôle.

A partir de ces deux remarques on voit bien ce qui se passe pour (Pa) si θ change de signe ou ne vérifie pas (3.3). Aussi dans ce qui suit nous supposerons souvent, pour simplifier, $\theta > 0$

4. Existence et unicité du contrôle optimal dans le case général

Nous allons utiliser les idées et résultats du chapitre 3 pour démontrer des résultats d'existence assez généraux. On suppose a(x,t) régulière en (x,t); et on désigne par $a^{**}(x,t)$ la convexifiée de a par rapport à t, à x fixé.

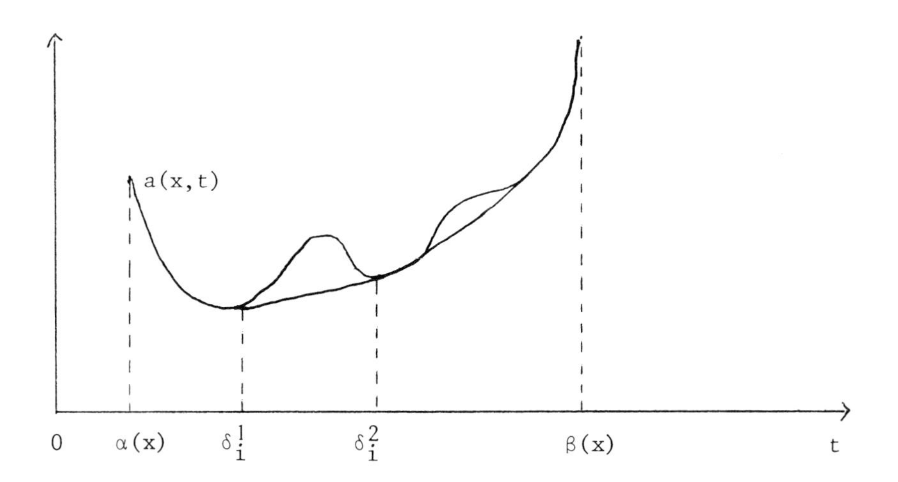

Figure 1.

Suivant la figure 1, notons

$$\theta_i(x) = \frac{a(x,\delta_i^2) - a(x,\delta_i^1)}{\delta_i^2 - \delta_i^1} \text{ p.p. } x \in \Omega_i = \{x / \delta_i^1(x) < \delta_i^2(x)\}$$

l'indice i appartenant à un ensemble I supposé dénombrable; et faisons l'hypothèse de type (3.3) :

$$\forall \lambda \in [\lambda_1(\nu_2), \lambda_1(\nu_1)] ,$$

$$\frac{-\Delta(1/\sqrt{\theta_i(x)})}{\lambda.(1/\sqrt{\theta_i(x)})} \notin \text{Im } a(x,.) \text{ p.p. } x \in \Omega_i \ \forall i \in I, \tag{4.1}$$

dans laquelle θ_i est supposée suffisamment régulière.

REMARQUE 4.1. Si α,β et a sont indépendants de x, l'hypothèse (4.1) est automatiquement satisfaite.
Nous avons le

THEOREME 4.1. Sous l'hypothèse (4.1) le problème (Pa) admet une solution optimale $(\bar{\omega},\bar{u})$.

Démonstration
Nous indiquons seulement les grandes lignes de la preuve. Le problème (Pa**) admet une solution optimale $(\bar{\omega},\bar{u})$; et nous pouvons montrer que cette solution est caractérisée par la condition d'optimalité

$$\int_\Omega \frac{\partial a^{**}}{\partial t} (x,\bar{u}).(v-\bar{u})\bar{\omega}^{-2}(x)dx \geqslant 0 \ \forall v \in U; \tag{4.2}$$

alors cette relation et l'hypothèse (4.1) permettent, en procédant comme dans la preuve du théorème 3.1, de montrer que l'ensemble

$$\{x \in \Omega / a^{**}(x,\bar{u}(x)) < a(x,\bar{u}(x))\}$$

est de mesure nulle i.e. que

$$a^{**}(x,\bar{u}(x)) = a(x,\bar{u}(x)) \text{ p.p. } x \in \Omega ,$$

quitte à modifier éventuellement $\bar{u}$ sur les ensembles de mesure non nulle où θ_i vaut zéro.

Ceci prouve que $(\bar{\omega},\bar{u})$ est aussi solution optimale de (Pa). □

REMARQUES 4.2.

1) Signalons qu'en fait, à travers la preuve du théorème 4.1, on montre que presque partout en x le couple $(\bar{u}(x), a(x,\bar{u}(x)))$ n'appartient pas aux parties affines de pente non nulle du graphe de $a(x,.)$

2) dans le cas unidimensionnel pour α,β constants et a indépendant de x le résultat ci-dessus a été obtenu dans [ZE].

COROLLAIRE 4.1. Avec l'hypothèse (4.1) les deux problèmes (Pa) et (Pa**) sont identiques si on a (4.2)' :

$$|\{x\in\Omega/\theta_i(x) = 0\}| = 0 \ \forall i\in I. \qquad (4.2)'$$

Démonstration

i) Le théorème (4.1), joint à (4.2)', montre que toute solution de (Pa**) est aussi solution de (Pa).

ii) Montrons la réciproque. Soit $(\omega,\tilde{u})$ une solution de (Pa); et supposons que l'ensemble

$$E = \{x\in\Omega/a^{**}(x,\tilde{u}(x)) < a(x,\tilde{u}(x))\}$$

soit de mesure non nulle. La première valeur propre $\lambda_1(a^{**},\tilde{u})$ de

$$\begin{cases} -\Delta\tilde{\omega} = \lambda_1(a^{**},\tilde{u}).a^{**}(x,\tilde{u}).\tilde{\omega} \text{ dans } \Omega \\ \tilde{\omega}/\Gamma = 0, \quad \tilde{\omega}>0 \quad \text{dans} \quad \Omega \end{cases}$$

vérifie l'inégalité suivante

$$\lambda_1(a^{**},\tilde{u}) = \frac{\int_\Omega |\nabla\tilde{\omega}|^2\,dx}{\int_\Omega a^{**}(x,\tilde{u})\tilde{\omega}^2dx} > \frac{\int_\Omega |\nabla\tilde{\omega}|^2\,dx}{\int_\Omega a(x,\tilde{u})\tilde{\omega}^2dx} \geq \inf_\varphi \frac{\int_\Omega |\nabla\varphi|^2dx}{\int_\Omega a(x,\tilde{u})\varphi^2dx} = \lambda_1(a,\tilde{u})$$

comme

$$\sup \{\lambda_1(a,v), v \in U\} = \sup \{\lambda_1(a^{**},v), v \in U\},$$

on a

$$\sup \{\lambda_1(a,v),\ v \in U\} > \lambda_1(a,\tilde{u}).$$

Ce qui constitue une contradiction. Et donc $(\omega,\tilde{u})$ est solution optimale de (Pa^{**}). □

Examinons maintenant l'unicité de la soltuion de (Pa). Ecartons tout de suite le cas où il n'y a pas unicité en général i.e. lorsque a^{**} possède un palier. Alors nous avons le resultat suivant

<u>THEOREME 4.2.</u> Supposons a, α et β indépendants de x, a^{**} sans palier. Alors (Pa) admet une solution optimale unique $(\bar{u},\bar{\omega})$, $\bar{\omega}$ étant déterminé à une normalisation près.

<u>Démonstration</u>

Dans le seul but de simplifier les notations et la présentation de la preuve nous supposerons que l'ensemble

$$K = \{ t \in [\alpha,\beta] \ / \ a^{**}(t) = \text{affine}\}$$

est convexe - (cf Figure 2)-. Le cas général s'en déduit sans difficulté.

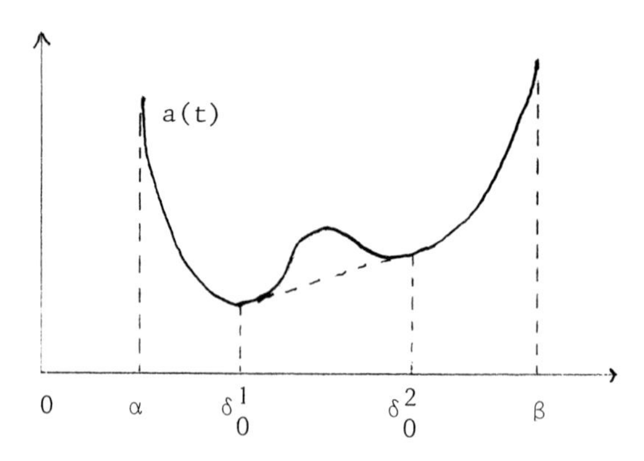

Figure 2.

Supposons qu'il existe deux solutions optimales (ω_1,u_1), (ω_2,u_2) telles que l'ensemble $E = \{x\in\Omega/u_1(x) \neq u_2(x)\}$ soit de mesure non nulle. Nous aurons besoin de deux étapes.

1^{ière} étape : démontrons le résultat suivant

<u>LEMME 4.1.</u> On a $\overset{-1}{u_1}(I_j) = \overset{-1}{u_2}(I_j)$ p.p.$x\in\Omega$, $j = 1,2$ où $I_1 = [\alpha,\delta_0^1]$, $I_2 = [\delta_0^2,\beta]$.

<u>Démonstration du lemme 4.1.</u>

D'après le théorème (4.1) u_i vérifie

$$u_i(x) \in I_1 \cup I_2 \text{ p.p.}x\in\Omega,\ \forall i = 1,2; \tag{4.3}$$

d'autre part on peut montrer que l'ensemble des contrôles optimaux de (Pa) est convexe; et par conséquent quel que soit θ dans $[0,1]$ on a :

$$u_\theta(x) = \theta u_1(x) + (1-\theta)u_2(x) \in I_1 \cup I_2 \text{ p.p.}x \in \Omega \tag{4.4}$$

Raisonnons alors par l'absurde i.e. supposons que l'ensemble

$$F = \overset{-1}{u_1}(I_1) \cap \overset{-1}{u_2}(I_2)$$

est de mesure strictement positive.

D'après le théorème de Lusin il existe un compact $K\subset F$, de mesure strictement positive, tel que $\tilde{u}_1 = u_1/K$ et $\tilde{u}_2 = u_2/K$ soient continues. Et alors la fonction

$$v : (\theta,x)\in[0,1]\times K \to v(\theta,x)\in\mathbb{R}^+$$

définie par

$$v(\theta,x) = \theta\tilde{u}_1(x) + (1-\theta)\tilde{u}_2(x)\ \forall(\theta,x)\in[0,1]\times K,$$

vérifie d'après (4.4)

$$v([0,1]\times K)\subseteq I_1 \cup I_2 \tag{4.5}$$

i.e.

$$\theta\tilde{u}_1(x) + (1-\theta)\tilde{u}_2(x) \in I_1 \cup I_2 \quad \forall x \in K, \ \forall\theta \in [\,0,1\,]. \tag{4.6}$$

Mais tout x dans K on a

$$u_i(x) \in I_i \quad i = 1,2,$$

d'après la définition de F. Ainsi tout $x_0 \in K$ il existe $\theta_0 = \theta(x_o) \in [\,0,1\,]$ tel que

$$v(\theta_o, x_o) \in]\,\delta_0^1 \ \delta_0^2[\ ;$$

ceci contredit (4.5); et finalement $|F| = 0$, ce qui termine la preuve du lemme . □

2ième étape : il s'agit de montrer que $u_1 = u_2$ ou en d'autres termes $|E| = 0$
Le contrôle $\bar{u} = \frac{1}{2}(u_1 + u_2)$ est optimal pour (Pa**); et il vérifie :

$$\frac{1}{\lambda_1(\bar{u})} = \frac{\int_\Omega a^{**}(\bar{u})\bar{\omega}^2 dx}{\int_\Omega |\nabla\bar{\omega}|^2 \, dx} = \sup_{\varphi\in H_0^1} \frac{\int_\Omega a^{**}(\bar{u})\varphi^2 \, dx}{\int_\Omega |\nabla\varphi|^2 \, dx} \tag{4.7}$$

où $\bar{\omega}$ est l'état optimal correspondant. La convexité de a**, le lemme 4.1 et la remarque (4.2) permettent d'écrire :

$$a^{**}(\bar{u}(x)) < \frac{1}{2}\, a^{**}(u_1(x)) + \frac{1}{2}\, a^{**}(u_2(x)) \ \text{p.p.} x \in E. \tag{4.8}$$

Enfin (4.7) et (4.8) entrainent (puisque par hypothèse $|E| > 0$)

$$\frac{1}{\lambda_1(\bar{u})} < \frac{1}{2}\left(\frac{1}{\lambda_1(u_1)} + \frac{1}{\lambda_1(u_2)}\right) = \frac{1}{\sup\lambda_1(v)},$$

i.e.

$$\lambda_1(\bar{u}) > \sup\{\lambda_1(v),\ v\in U\} = \lambda_1(u_1) = \lambda_1(u_2)$$

contredisant ainsi l'optimalité de u_1 et u_2. Donc $|E| = 0$; ceci achève la démonstration du théorème. □

REMARQUE 4.3. Dans le cas dépendant de x, l'unicité reste posée.

5. Condition nécessaire d'existence de contrôle optimal bang-bang.

Les hypothèses (3.1) et (3.2) sont-elles nécessaires pour que le problème (Pa) admette un contrôle optimal bang-bang ?
Dans le cas où α, β et a sont indépendants de x nous avons le résultat suivant

THEOREME 5.1. Supposons α, β, a indépendants de x et a dérivable. Alors les hypothèses (3.1) et (3.2) expriment une condition necessaire et suffisante d'existence d'un contrôle optimal bang-bang.

Démonstration

Désignons par a^{**} la convexifiée de a.

i) (3.1) et (3.2) forment une condition suffisante d'existence d'un contrôle optimal bang-bang : ceci est donné par le théorème 3.1.

ii) La condition nécessaire : Ecartons le cas $\theta = \dfrac{a(\beta) - a(\alpha)}{\beta-\alpha} = 0$ qui est immédiat.

Nous ferons un raisonnement par l'absurde. Rappelons que toute solution de (Pa) est également solution de (Pa**) . Par conséquent il suffit de montrer que si l'hypothèse (3.1) - (3.2) n'a pas lieu, (Pa**) ne peut admettre de contrôle optimal possédant la propriété bang-bang. Pour cela supposons que l'on ait :

$$\frac{da^{**}}{dt}(\alpha) < \frac{da^{**}}{dt}(\beta) \tag{5.1}$$

et que (Pa**) possède un contrôle optimal $\bar{u}$ bang-bang; soit $\bar{\omega}$ l'état correspondant; $(\bar{\omega},\bar{u})$ vérifie la relation d'optimalité

$$\int_\Omega \frac{da^{**}}{dt}(\bar{u}).(v-\bar{u})\bar{\omega}^{2}\, dx \geqslant 0 \ \forall\ v \in U, \tag{5.2}$$

que l'on peut écrire aussi sous la forme :

$$[\frac{da^{**}}{dt}(\bar{u}(x)).\bar{\omega}^{2}(x) + \mu]\, v \geqslant [\frac{da^{**}}{dt}(\bar{u}(x)).\bar{\omega}^{2}(x) + \mu]\, \bar{u}(x)$$

$$\text{p.p.}\, x \in \Omega,\ \forall v,\ \alpha \leqslant v \leqslant \beta \tag{5.3}$$

où le réel μ désigne un multiplicateur de Lagrange. Comme $\bar{u}$ est bang-bang posant

$$\Omega_\alpha = \{x \in \Omega / u(x) = \alpha\},$$

$$\Omega_\beta = \{x \in \Omega / \bar{u}(x) = \beta\},$$

nous avons

$$\Omega = \Omega_\alpha \cup \Omega_\beta\ ;$$

et en prenant $v = \beta$, (5.3) s'écrit dans Ω_α

$$\frac{da^{**}}{dt}(\alpha)\bar{\omega}^{2}(x) + \mu \geqslant 0\ \text{p.p.}\, x \in \Omega_\alpha. \tag{5.4}$$

On se donne alors δ arbitraire tel que

$$\frac{da^{**}}{dt}(\alpha) < \delta < \frac{da^{**}}{dt}(\beta),$$

et on majore (5.4) :

$$\delta\bar{\omega}^{2}(x) + \mu \geqslant 0\ \text{p.p.}\, x \in \Omega_\alpha; \tag{5.5}$$

de même nous montrons que

$$\delta\bar{\omega}^{2}(x) + \mu \leqslant 0\ \text{p.p.}\, x \in \Omega_\beta. \tag{5.6}$$

Mais on peut montrer qu'il existe $t_o > 0$ tel que

$$\Omega_\alpha = \{x \in \Omega / \bar{\omega}(x) > t_o\},$$

$$\Omega_\beta = \{x \in \Omega / \bar{\omega}(x) < t_o\},$$

à un ensemble de mesure nulle près. Soit alors x_o appartenant à $\{x \in \Omega / \bar{\omega}(x) = t_o\}$ tel qu'il existe $x_n \in \Omega_\alpha$, $y_n \in \Omega_\beta$ satisfaisant :

$$x_n \to x_o \quad ,$$

$$y_n \to y_o .$$

Comme $\bar{\omega}$ est continue, par passage à la limite nous obtenons à partir de (5.5) et (5.6) les inégalités suivantes :

$$\delta \bar{\omega}^2(x_o) + \mu \geqslant 0$$

et

$$\delta \bar{\omega}^2(x_o) + \mu \leqslant 0$$

pour tout δ dans $] \frac{da^{**}}{dt}(\alpha), \frac{da^{**}}{dt}(\beta) [$; ceci constitue une contradiction. Ainsi $\bar{u}$ ne peut être bang-bang. □

Dans le cas où α, β et a dépendent de x, que peut-on dire de la condition nécessaire d'existence d'un contrôle optimal bang-bang.

Dans cette situation nous ne disposons que d'une réponse partielle. Introduisons quelques notations et définitions. Soit la mesure positive μ définie par :

$$\mu(E) = \int_E (\beta - \alpha) dx$$

pour tout mesurable $E \subseteq \Omega$; notons

$$\xi_1 = \int_\Omega \beta \, dx - \gamma,$$

$$\xi_2 = \gamma - \int_\Omega \alpha \, dx \ ,$$

$$S_i = \sup \{ |E| / E \text{ mesurable} \subseteq \Omega, \ \mu(E) = \xi_i \} \quad i = 1,2.$$

REMARQUE 5.1. [RU]. Rappelons pour mémoire que le lemme classique de Lyapounov entraine l'existence d'un mesurable $D_i \subseteq \Omega$ tel que

$$|D_i| = S_i, \ i = 1,2.$$

Nous nous donnons trois fonctions $\alpha(x)$, $\beta(x)$, $a(x,t)$ continues, a dérivable en t telles que l'ensemble ouvert :

$$\tilde{\Omega} = \{x \in \Omega / \frac{\partial a^{**}}{\partial t}(x,\alpha) < \frac{\partial a^{**}}{\partial t}(x,\beta)\}$$

soit connexe et vérifie :

$$|\tilde{\Omega}| > \max \{S_1, S_2\}. \tag{5.7}$$

Il est clair que (5.7) entraine que l'hypothèse (3.1) - (3.2) n'est pas vérifiée; pour simplifier l'énoncé on suppose $\theta = \frac{a(x,\beta) - a(x,\alpha)}{\beta - \alpha} > 0$. On a

<u>THEOREME 5.2</u>. On suppose que l'on a (4.1), (3.3), (5.7) et $\theta > 0$. Alors tout contrôle optimal $\bar{u}$ de (Pa) vérifie

$$|\{x \in \Omega / \alpha(x) < \bar{u}(x) < \beta(x)\}| > 0$$

i.e. $\bar{u}$ ne peut être bang-bang.

<u>Démonstration</u> :

D'après le théorème (4.1) (Pa) admet au moins une solution.
Nous raisonnerons par l'absurde i.e. soit $(\bar{\omega}, \bar{u})$ une solution optimale de (Pa) telle que $\bar{u}$ possède la propriété bang-bang; $(\bar{\omega}, \bar{u})$ est aussi solution de (Pa**); et on montre facilement que $(\bar{\omega}, \bar{u})$ est également solution optimale de (P_L) avec

$$L(x,t) = \theta(x).t + \tilde{s}(x),$$

$$\tilde{s}(x) = \frac{\beta a(x,\alpha) - \alpha\, a(x,\beta)}{\beta - \alpha}.$$

Par conséquent, compte tenu de (4.1), il existe, d'après le théorème (3.1) un réel $t_o > 0$ tel que

$$\bar{u} = \begin{cases} \alpha \text{ sur } \Omega_\alpha , \\ \beta \text{ sur } \Omega_\beta = \Omega \setminus \Omega_\alpha \end{cases}$$

où

$$\Omega_\alpha = \{x \in \Omega / \theta(x)\bar{\omega}^{-2}(x) > t_o\} .$$

L'hypothèse (5.7) nous permet d'affirmer que

$$|\Omega_\alpha \cap \tilde{\Omega}| > 0,$$

$$|\Omega_\beta \cap \tilde{\Omega}| > 0$$

car

$$\mu(\Omega_\alpha) = \xi_1 \quad \text{et } \mu(\Omega_\beta) = \xi_2.$$

Par ailleurs il est possible de choisir un ouvert Ω' connexe tel que

$$\bar{\Omega}' \subset \tilde{\Omega}$$

et de sorte que les deux ensembles

$$\Omega'_\alpha = \Omega_\alpha \cap \Omega',$$

$$\Omega'_\beta = \Omega_\beta \cap \Omega'$$

soient de mesure strictement positive. Mais le couple $(\bar{\omega},\bar{u})$ est également solution de (Pa**); par conséquent il vérifie la relation d'optimalité suivante

$$\int_{\Omega'} \frac{\partial a^{**}}{\partial t}(x,\bar{u}).(v-\bar{u})\bar{\omega}^{-2}dx \geqslant 0 \tag{5.8}$$

pour tout v dans $L^\infty(\Omega')$ satisfaisant

$$\alpha \leqslant v \leqslant \beta, \int_{\Omega'} v \, dx = \gamma - \int_{\Omega\setminus\Omega'} \bar{u} \, dx ;$$

(5.8) peut s'écrire sous la forme

$$\begin{aligned} &\frac{\partial a^{**}}{\partial t}(x,\alpha)\bar{\omega}^{-2}(x) + \mu \geqslant 0 \text{ p.p.} x \in \Omega'_\alpha \quad , \\ &\frac{\partial a^{**}}{\partial t}(x,\beta)\bar{\omega}^{-2}(x) + \mu \leqslant 0 \text{ p.p.} x \in \Omega'_\beta \quad , \end{aligned} \tag{5.9}$$

où le réel μ est un multiplicateur de Lagrange. Pour conclure on se propose de majorer $(5.9)_1$, minorer $(5.9)_2$ et d'aboutir ainsi à une contradiction entre les deux inégalités obtenues après intégration sur des ensembles convenables notés Δ_n^α, Δ_n^β que nous définirons en (5.11) et (5.15). Pour cela posons

$$s(x) = \frac{\theta(x) + a^{**\prime}(x,\alpha(x))}{2} \quad \text{sur } \tilde{\Omega}$$

où

$$a^{**\prime}(x,t) = \frac{\partial a^{**}}{\partial t}(x,t),$$

$$\sigma(x) = \theta(x) - s(x) \text{ sur } \tilde{\Omega} ,$$

$$k_o = \inf \{\sigma(x),\ x \in \bar{\Omega}'\};$$

nous avons $k_o > 0$ puisque $\sigma(x) > 0$ sur $\bar{\Omega}'$. Considérons les majorations suivantes de $(5.9)_1$

$$\begin{aligned} 0 &\leqslant a^{**\prime}(x,\alpha(x))\bar{\omega}^{-2}(x) + \mu \leqslant s(x)\bar{\omega}^{-2}(x) + \mu \\ &< s(x)\bar{\omega}^{-2}(x) + \frac{k_o}{2}\bar{\omega}^{-2}(x) + \mu < \theta(x)\bar{\omega}^{-2}(x) + \mu \text{ p.p.} x \in \Omega'_\alpha \end{aligned} \tag{5.10}$$

dont nous allons prendre les moyennes sur l'ensemble mentionné plus haut

$$\Delta_n^\alpha = \{x \in \Omega' / t_o + \frac{1}{n} \geqslant \theta(x)\bar{\omega}^{-2}(x) \geqslant t_o + \frac{1}{n^2}\} \tag{5.11}$$

qui est inclus dans Ω'_α . Pour ce faire il nous faut montrer que $|\Delta_n^\alpha| > 0$. Et ceci résulte, d'une part de la continuité de $\theta\bar{\omega}^{-2}$ et d'autre part de

$$|\{x \in \Omega' / \theta(x)\bar{\omega}^{-2}(x) \leqslant t_o + \frac{1}{n^2}\}| > 0 \tag{5.12}$$

et

$$|\{x \in \Omega' / \theta(x)\bar{\omega}^{-2}(x) \geqslant t_o + \frac{1}{n}\}| > 0 \tag{5.13}$$

qui vient du fait que Ω' est connexe et que

$$|\{x \in \Omega' / \theta\bar{\omega}^{-2}(x) > to\}| = |\{x \in \Omega' / \theta\bar{\omega}^{-2}(x) \geqslant to\}| > 0$$

-(pour un résultat général nous renvoyons à [CH2])-. Donc nous avons

$$0 \leq I_\alpha < I_\alpha + \frac{k_o}{2} \lim_n \frac{1}{|\Delta_n^\alpha|} \int_{\Delta_n^\alpha} \bar{\omega}^{-2}(x)dx \leq to + \mu,$$

où

$$I_\alpha = \lim_n \frac{1}{|\Delta_n^\alpha|} \int_{\Delta_n^\alpha} s\bar{\omega}^{-2}dx + \mu \ ;$$

ceci donne

$$to + \mu > 0 \tag{5.14}$$

Enfin pour achever la preuve du théorème il suffit de minorer $(5.9)_2$ au sens large et de prendre sa moyenne sur

$$\Delta_n^\beta = \{x \in \Omega' / to - \frac{1}{n} \leq \theta\bar{\omega}^{-2}(x) \leq t \ - \frac{1}{n^2}\} \tag{5.15}$$

dont la mesure est strictement positive pour les mêmes raisons que dans le cas Δ_n^α; ainsi nous obtenons

$$\mu + \lim_n \frac{1}{|\Delta_n^\beta|} \int_{\Delta_n^\beta} \theta\, \bar{\omega}^{-2}dx \leq \mu + \lim_n \frac{1}{|\Delta_n^\beta|} \int_{\Delta_n^\beta} a^{**\prime}(x,\beta)\bar{\omega}^{-2}dx \leq 0$$

i.e.

$$to + \mu \leq 0;$$

et cette dernière inégalité contredit (5.14). c.q.f.d. □

<u>REMARQUE 5.2</u> . L'hypothèse de connexité de $\tilde{\Omega}$ ne semble pas naturelle; il ne devrait intervenir que l'hypothèse (5.7) portant sur la taille de $\tilde{\Omega}$. C'est par exemple le cas si $\tilde{\Omega}$ est donné par

$$\tilde{\Omega} = \{x \in \Omega / a^{**\prime}(x,\alpha(x)) \leq 0\}$$

et satisfait (5.7). Dans ce cas on procède différemment : on exhibe un contrôle admissible $v_0 = v(\tilde{\Omega})$ contredisant (5.2) si le contrôle optimal $\bar{u}$ est bang-bang.

6. Un ensemble général de contrôles admissibles

Dans toute cette section on choisit α et β constants et $a(t) = t$. Pour toute fonction h mesurable de Ω dans $\mathbb{R}$ rappelons les définitions classiques suivantes [HLP],[CR],[MO].

DEFINITIONS 6.1.

1) La fonction de distribution de h est la fonction δ_h de $\mathbb{R}$ dans $[0,|\Omega|]$, définie par :

$$\delta_h(t) = |\{x \in \Omega / h(x) < t\}|.$$

2) Le réarrangement croissant (unidimensionnel) de h est la fonction notée h_* de $[0,|\Omega|]$ dans $\overline{\mathbb{R}}$ telle que

$$\begin{aligned} h_*(s) &= \inf \{t \in \mathbb{R} / \delta_h(t) > s\} \quad \text{si } s \in [0,|\Omega|[\\ h_*(|\Omega|) &= \sup_{\Omega} \text{ess } h \end{aligned} \qquad (6.1)$$

3) Le réarrangement décroissant (unidimensionnel) de h est défini par :

$$h^*(s) = h_*(|\Omega| - s) \text{ p.p.s} \qquad (6.2)$$

4) On appelle ensemble de niveau ou ensembles équipotentiels de h tout sous-ensemble de Ω de type

$$E_1(t) = \{x \in \Omega / h(x) > t\}, \ t \in \mathbb{R}$$

ou

$$E_2(t) = \{x \in \Omega / h(x) < t\}, \ t \in \mathbb{R}$$

avec éventuellement des inégalités au sens large. Et nous désignerons par "courbes de niveau" de h les ensembles tel que

$$E_3(t) = \{x \in \Omega / h(x) = t\}, \ t \in \mathbb{R}.$$

Etant donnée une fonction croissante u_o définie sur $[0,|\Omega|]$, telle que

$$0 < \alpha \leq u_o(s) \leq \beta \text{ p.p.} s \in [0,|\Omega|],$$

on introduit, à la place de (2.3), l'ensemble des contrôles admissibles suivant :

$$U = \{v \in L^{\infty}(\Omega)/\alpha \leq v \leq \beta,\ v_* = u_o\}^{(1)}. \tag{6.3}$$

Et on se propose de résoudre le problème (Pa) qui, rappelons-le consiste à trouver $\bar{u}$ dans U satisfaisant

$$\lambda_1(\bar{u}) = \sup \{\lambda_1(v),\ v \in U\}$$

où $\lambda_1(v)$ est la première valeur propre de

$$\begin{cases} -\Delta\omega = \lambda_1(v).v.\omega \\ \omega/\Gamma = 0,\ \omega > 0. \end{cases} \tag{6.4}$$

Nous avons le résultat suivant

THEOREME 6.1. (Pa) admet une solution optimale $(\bar{\omega},\bar{u})$. Cette solution vérifie les propriétés suivantes :

1) $\bar{\omega}$ et $\bar{u}$ ont "les mêmes ensembles équipotentiels" au sens suivant : tout $t > 0$ il existe $s = s(t)$ tel que :

$$\{x \in \Omega/\bar{u}(x) < s\} \subseteq \{x \in \Omega/\bar{\omega}(x) \geq t\} \subseteq \{x \in \Omega/\bar{u}(x) \leq s\}.$$

2) $\bar{u}$ est fonction décroissante de $\bar{\omega}$; de façon plus précise $\bar{u}$ a pour expression :

$$\bar{u}(x) = \phi(\bar{\omega}(x)) \text{ p.p. } x \in \Omega$$

où

$$\phi(t) = u_o((\bar{\omega}^*)^{-1})(t)$$

3) $$\int_\Omega \bar{u}.\bar{\omega}^{-2}dx = \int_0^{|\Omega|} (\bar{u})_*.(\bar{\omega}^*)^2 ds.$$

Démonstration

L'ensemble U étant non convexe, on se propose de relaxer le problème en

(1) Des ensembles de type (6.3) sont également considérés dans [ALT1].

remplaçant U par

$$\tilde{U} = \{v \in L^\infty(\Omega) / \alpha \leqslant v \leqslant \beta, \int_0^t v_*(s)ds \geqslant \int_0^t u_0(s)ds, \forall t \in [0,|\Omega|],$$

et

$$\int_0^{|\Omega|} v_*(s)ds = \int_0^{|\Omega|} u_o(s)ds\};$$

cet ensemble est convexe puisqu'il s'écrit :

$$\tilde{U} = \{v \in L^\infty(\Omega) / \alpha \leqslant v \leqslant \beta, \int_E v(x)dx \geqslant \int_0^{|E|} u_o(s)ds,$$

$$\forall \text{ mesurable } E \subseteq \Omega, \text{ et } \int_\Omega v(x)dx = \int_0^{|\Omega|} u_o(s)ds\}.$$

D'après le résultat de [MI1] $\tilde{U}$ est la fermeture de U pour la topologie $\sigma(L^\infty, L^1)$. Le problème relaxé, noté $(\tilde{Pa})$, qui s'écrit

$$(\tilde{Pa}) \begin{cases} -\Delta\omega = \lambda_1(v).v.\omega & \text{dans } \Omega, \\ \omega/_{\partial\Omega} = 0, \ \omega > 0 & \text{dans } \Omega, \\ \sup \{\lambda_1(v), \ v \in \tilde{U}\}, & \end{cases}$$

admet donc une solution optimale $(\bar{\omega},\bar{u})$; et la première partie du théorème 6.1 sera prouvée si on montre que $\bar{u}$ appartient à $U \subset \tilde{U}$ (étape 1); l'étape 2 sera consacrée aux informations sur $\bar{u}$.

1ière étape :

Montrons que $\bar{u} \in U$. Pour ce faire on définit

$$\phi(t) = u_0((\bar{\omega}^*)^{-1})(t);$$

on part de la relation d'extrémalité satisfaite par $(\bar{\omega},\bar{u})$

$$\int_\Omega v\bar{\omega}^2 dx \geqslant \int_\Omega \bar{u}\,\bar{\omega}^2 dx \ \forall v \in \tilde{U}, \tag{6.5}$$

et de l'inégalité

$$A(t) = \int_0^{t} \bar{u}_*(s)ds \geq \int_0^{t} u_0(s)ds = A_0(t),$$

pour laquelle l'égalité est atteinte lorsque $t = |\Omega|$, puisque $\bar{u} \in \tilde{U}$. A partir de l'inégalité de Hardy-Littlewood

$$\int_\Omega \bar{u}.\bar{\omega}^2 dx \geq \int_0^{|\Omega|} \bar{u}_*(\bar{\omega}^*)^2 ds, \tag{6.6}$$

on fait deux intégrations par parties en introduisant successivement A(t) et $A_0(t)$ à partir du second membre de (6.6); utilisant la définition de ϕ, on obtient après quelques calculs la série d'inégalités et d'égalités suivantes

$$\int_\Omega \bar{u}\bar{\omega}^2 dx \geq \int_0^{|\Omega|} \bar{u}_*(\bar{\omega}^*)^2 dx \geq \int_0^{|\Omega|} u_0(\bar{\omega}^*)^2 ds = \int_0^{|\Omega|} \phi(\bar{\omega}^*).(\bar{\omega}^*)^2 dx$$

$$= \int_\Omega \phi(\bar{\omega}).\bar{\omega}^2 dx, \tag{6.7}$$

la dernière égalité étant obtenue par équimesurabilité. Dans (6.5) nous prenons pour v le contrôle admissible :

$$v = \phi(\bar{\omega}),$$

et après comparaison de (6.5) et (6.7), nous obtenons les égalités

$$\int_\Omega \bar{u}\,\bar{\omega}^2 dx = \int_0^{|\Omega|} \bar{u}_*(\bar{\omega}^*)^2 ds = \int_0^{|\Omega|} u_0(\omega^*)^2 ds = \int_\Omega \phi(\bar{\omega})\bar{\omega}^2 dx \tag{6.8}.$$

Après quelques calculs on peut montrer à l'aide de la deuxième égalité (6.8) que

$$\bar{u}_*(s) = u_o(s) \text{ p.p.s};$$

ceci entraine que $\bar{u} \in U$ i.e. $(\bar{\omega}, \bar{u})$ est solution de (Pa); ainsi s'achève l'étape 1.

$2^{\underline{e}}$ étape :

A partir de l'égalité

$$\int_\Omega \bar{u}(x).\bar{\omega}^2(x)dx = \int_0^{|\Omega|} \bar{u}_*(s)(\bar{\omega}^*(s))^2 ds,$$

utilisant la formule -(qu'on obtient comme dans [TAL2], [MO])-

$$\frac{d}{dt}\int_{E(t)} \bar{u}(\bar{\omega} - t)dx = -\int_{E(t)} \bar{u}dx \text{ p.p.t}$$

dans laquelle $E(t) = \{x\in\Omega/\bar{\omega}(x) < t\}$, nous pouvons montrer après quelques calculs que pour tout $t>0$ il existe $s = s(t)$ tel que ([TA1]).

$$\{x\in\Omega/\bar{u}(x) < s(t)\} \subseteq \{x\in\Omega/\bar{\omega}(x) \geqslant t\} \subseteq \{x\in\Omega/\bar{u}(x) \leqslant s(t)\} \ . \qquad (6.9)$$

Enfin $\bar{u}$ est fonction décroissante de $\bar{\omega}$ puisque d'après la définition de $\phi(t)$ et le fait que $\bar{u}\in U$ on a :

$$\bar{u}(x) = \phi(\bar{\omega}(x))$$ □

REMARQUE 6.1. Si u_o est strictement croissante (6.9) entraine que pour tout $t>0$

$$\{x\in\Omega/\bar{\omega}(x) > t\} = \{x\in\Omega/\bar{u}(x) < s(t)\} \text{ p.p.}$$

Examinons maintenant l'unicité de la solution optimale; nous avons le résultat suivant :

THEOREME 6.2. La solution $(\bar{u},\bar{\omega})$ de (Pa) est unique.

Démonstration .

Elle repose sur la propriété 3) de la solution optimale. Supposons qu'il existe deux solutions de (Pa) $(\bar{\omega},\bar{u})$ et $(\underline{\omega},\underline{u})$, $\bar{\omega}$ et $\underline{\omega}$ étant définis à une de normalisation près; ces solutions vérifient la série d'égalités (6.8) et donc

$$\int_\Omega \bar{u}\,\bar{\omega}^2 dx = \int_0^{|\Omega|} \bar{u}_* (\bar{\omega}^*)^2 ds$$

$$\int_\Omega \underline{u}\underline{\omega}^2\, dx = \int_0^{|\Omega|} (\underline{u})_* .(\underline{\omega}^*)^2 ds$$

Comme

$$\bar{u}_* = (\underline{u})_* = u_0,$$

on a

$$\int_\Omega \bar{u}\, \bar{\omega}^2 dx = \int_0^{|\Omega|} (\underline{u})_* . (\bar{\omega}^*)^2 ds$$

$$\int_\Omega \underline{u}\, \underline{\omega}^2 dx = \int_0^{|\Omega|} \bar{u}_* \, . (\underline{\omega}^*)^2 ds; \qquad (6.10)$$

qui entraine les deux égalités

$$\int_0^{|\Omega|} (\underline{u})_* . (\bar{\omega}^*)^2 dx = \int \underline{u}\bar{\omega}^2 dx,$$

$$\int_0^{|\Omega|} (\bar{u})_* . (\underline{\omega}^*)^2 ds = \int_\Omega \bar{u}\underline{\omega}^2 dx; \qquad (6.11)$$

en effet une inégalité en $(6.11)_1$ permettrait d'avoir

$$\lambda_1(\bar{u}) > \frac{\int_\Omega |\nabla\bar{\omega}|^2 dx}{\int_\Omega \underline{u}\bar{\omega}^2 dx} \geq \lambda_1(\underline{u})$$

ce qui contredirait l'optimalité de $\underline{u}$. De même si on avait une inégalité en $(6.11)_2$, on aurait $\lambda_1(\underline{u}) > \lambda_1(\bar{u})$ qui est impossible. Alors (6.11) joint à (6.10) permet d'écrire

$$\int_\Omega \bar{u}\, \bar{\omega}^2 dx = \int_\Omega \underline{u}\, \bar{\omega}^2 dx,$$

et donc

$$\lambda_1(\bar{u}) = \frac{\int_\Omega |\nabla\bar{\omega}|^2 dx}{\int_\Omega \underline{u}\, \bar{\omega}^2 dx} \geq \inf_{\varphi \in H_0^1} \frac{\int_\Omega |\nabla\varphi|^2 dx}{\int_\Omega \underline{u}\, \varphi^2 dx} = \lambda_1(\underline{u})$$

i.e.

$$\lambda_1(\underline{u}) = \frac{\int_\Omega |\nabla\bar{\omega}|^2 dx}{\int_\Omega \underline{u}\bar{\omega}^2 dx} = \inf_\varphi \frac{\int_\Omega |\nabla\varphi|^2 dx}{\int_\Omega \underline{u}\varphi^2 dx} \qquad (6.12)$$

puisque $\lambda_1(\bar{u}) = \lambda_1(\underline{u})$; l'équation d'Euler de (6.12) :

$$\begin{cases} -\Delta\bar{\omega} = \lambda_1(\underline{u}).\underline{u}\bar{\omega} & \text{dans } \Omega \\ \bar{\omega}/_{\Gamma} = 0,\ \bar{\omega} > 0 & \text{dans } \Omega, \end{cases}$$

comparée à l'équation

$$\begin{cases} -\Delta\bar{\omega} = \lambda_1(\bar{u}).\bar{u}\ \bar{\omega} & \text{dans } \Omega \\ \bar{\omega}_{/\Gamma} = 0,\ \bar{\omega} > 0 & \text{dans } \Omega, \end{cases}$$

donne

$$\bar{u} = \underline{u} \quad \text{p.p.} x \in \Omega$$

puisque $\bar{\omega}(x) > 0$ dans Ω; et il s'en suit que $\bar{\omega} = \underline{\omega}$ à une normalisation près. D'où l'unicité. □

<u>Exemple</u> : Contrôle de domaine

Etant données trois constantes $0 < \alpha = \alpha_1 < \alpha_2 < \alpha_3 = \beta$, on considère l'ensemble des contrôles suivants

$$U = \{D = (D_1, D_2, D_3),\ \text{mesurable} \subset \Omega^3 / |D_i| = \gamma_i,\ D_i \cap D_j = \phi$$
$$\text{si } i \neq j,\ \bigcup_{i=1}^{3} D_i = \Omega\},$$

et on se pose le problème

$$(\pi_a) \begin{cases} -\Delta\omega_D = \lambda_1(D).v_D.\omega_D & \text{dans } \Omega\ , \\ \omega_D/_{\partial\Omega} = 0,\ \omega_D > 0 & \text{dans }\ , \\ \sup\ \{\lambda_1(D),\ D \in U\} \end{cases}$$

où l'on a posé

$$v_D(x) = \alpha_i \quad \text{si } x \in D_i,\ i = 1,2,3.$$

Ce problème admet une solution optimale unique ($\bar{\omega} = \omega_{D^o}$, $\bar{u} = u_{D^o}$); de plus il existe deux réels $t_1 > 0$ et $t_2 > 0$ tels que

$$\bar{u}(x) = \alpha_i \quad \text{si } x \in D_i^0$$

où $D^0 = (D_1^0, D_2^0, D_3^0)$, défini à un ensemble de mesure nulle près, est caractérisé par

$$D_1^o = \{x \in \Omega / \bar{\omega}(x) > t_1\}\ ,$$

$$D_2^o = \{x \in \Omega / t_1 \geqslant \bar{\omega}(x) > t_2\},$$

$$D_3^o = \{x \in \Omega / t_2 \geqslant \bar{\omega}(x)\}\ .$$

L'existence et l'unicité de la solution du problème ci-dessus s'obtient à l'aide des théorèmes 6.1. et 6.2. en remarquant que (π_a) se formule comme suit : si nous notons

$$u_0(s) = \begin{cases} \alpha_1 & \text{si} \quad 0 \leqslant s < \gamma_1 \\ \alpha_2 & \text{si} \quad \gamma_1 \leqslant s < \gamma_1 + \gamma_2 \\ \alpha_3 & \text{si} \quad \gamma_1 + \gamma_2 \leqslant s < \gamma_1 + \gamma_2 + \gamma_3 = |\Omega| \end{cases}$$

nous avons pour tout $D \in U$ $(v_D)_* = u_o$ par conséquent (π_a) est identique à (Pa) avec l'ensemble des contrôles donnés par (6.3).

<u>REMARQUE 6.2.</u> Dans ce cadre le résultat optimal est obtenu à l'aide de trois matériaux rangés dans l'ordre croissant de leur densité. Il existe des situationsoù les matériaux ne sont pas rangés en fonction de l'ordre de leur densité (G. Milton); [ALT] et celles où il y a mélange des matériaux [MT].

Signalons enfin que l'autres questions sur ce sujet seront traitées dans [TA1] et que l'on peut appliquer les idées de ce travail au contrôle dans les équations elliptiques [TA1,2] .

Je remercie vivement L. Tartar pour son aide et ses conseils. Egalement je remercie vivement F. Murat et D. Serre pour leurs remarques et les discussions que l'ai eu avec eux sur ce sujet.

References

[ALT1] A. Alvino - P.L. Lions - G. Trombetti. On optimization problems with prescribed rearrangements. Pub.Dipart.Mat.Appl.R.Caccioppoli n°31, (1986) Univers. Napoli.

[ALT2] A. Alvino - P.L. Lions - G. Trombetti, A remark on comparison results via symmetrization. Proc.Roy.Soc. Edim (1985).

[AT] A. Alvino, G. Trombetti, A Lower Bound for the First Eigenvalue of an Elliptic Operator. J. Math. Anal. Appl. 94,(1983), p.328-337.

[B] D.O. Banks, Bounds for the Eigenvalues of Nonhomogeneous Hinged Vibrating Rods. J. Math. Mech. Vol. 16, no. 9, (1967), p.949-966.

[CH1] Optimization of distributed parameter structures vol. I and II. Proceedings of the NATO IOWA U.S.A. May 20 - June 4 1980. Edit. J. Cea and E. Haug

[CH2] A. Chabi - A. Haraux, Un théorème des valeurs intermédiaires dans les espaces de Sobolev et applications Pub. Lab. An. Num. Université Paris VI, no 84014.

[CR] K.M. Chong, N.M. Rice, Equimeasurable Rearrangements of Functions Queen's Papers in Pure and Applied Mathematics no 28. Queen's University, Kingston, Ontario, Canada (1971).

[E] H. Egnell, Extremal proporties of the first eigenvalue of a class of elliptic eigenvalue problems. Uppsala University. Department of Mathematics. Report no 7, April(1985).

[FM] G.A. Francfort, F. Murat, Homogenization and optimal bounds in linear elasticity. Pub. Lab. An. Num. Université Paris VI, no. 85009.

[HLP] G.H. Hardy, J.E. Littlewood, G. Polya, Inequalities. 2nd ed., Cambridge (1952).

[HR] E. Haug - B. Rousselet, Proceedings IOWA (1980) vol. II p. 1371.

[J] C. Jouron, Analyse théorique et numérique de quelques problèmes d'optimisation intervenant en théorie des structures. Thèse de Doctorat d'Etat, Université Paris-Sud (1976).

[K] M.G. Krein, On certain problems of the maximum and minimum of characteristic values and Lyapunov zones of stability. Amer. Math. Soc. Translations ser. 2, 1, (1955), p. 163-187.

[L] P.L. Lions, Two Geometrical Properties of Solutions of Semilinear Problems. Appl. Anal. (1981), vol. 12, p. 267-272.

[MI1] L. Migliaccio, Sur une condition de Hardy, Littlewood,Polya C.R.A.S. Paris, t 297, Serie I, p. 25-28 (1983).

[MI2] F. Mignot, Contrôle de fonctions propres C.R.A.S. Paris, t 280, serie A, p. 333-335 (1975).

[MO] J. Mossino, Inégalités isopérimétriques et applications en physique. Hermann Edit. Collect. Travaux en cours.

[MP] G. Milton and N. Phan-Thien, New bounds on effective elastic moduli of two-component materials. Proc. R. Soc. Lond. A380, 305-331 (1982).

[MU1] F. Murat, Existence pour un problème du type Haug-Rousselet-Jouron par une méthode de H-convergence. Résultats non publiés Mai-Juin (1979).

[MU2] F. Murat, Control in coefficients. Publ. Lab. Anal. Num.Université Paris VI, no. 83014.

[MT] F. Murat - L. Tartar, Calcul des variations et homogéneisation. Publ. Lab. An. Num. Université Paris VI, no. 84012.

[R] B. Rousselet, Optimal Design and Eigenvalue Problems. Proceedings 8th IFIP Conference. Lect. Notes in cont. and Inf. Sciences, Springer Verlag (1978).

[RU] W. Rudin, Real and Complex Analysis. Mc Graw-Hill Edit.

[S] J. Sokolowski, Optimal Control in Coefficients of Boundary Value Problems with Unilateral Constraints. Préprint (1984).

[TA1] R. Tahraoui, Thèse Paris VI (1986).

[TA2] R. Tahraoui, Contrôle optimal à réarrangement fixé. C.R.Acad.Sc.Paris, t 303,Série I ,n°19,(1986) pp.955-958 et article à paraître.

[TA3] R. Tahraoui, Sur le contrôle des valeurs propres C.R.A.S. Paris, t 300, Série I, no 4, p.101-104 (1985).

[TAL1] G. Talenti, Estimates for Eigenvalues of Sturm-Liouville Problems. General Inequalities 4 (Edited by W. Walter), 341-350 Birkhäuser (1984).

[TAL2] G. Talenti, Elliptic equations and rearrangements, Ann. della Scuola Norm. Sup. di Pisa, Serie 4, no 3 (1976), p. 697-718.

[TAR1] L. Tartar, Exposé au Séminaire E.N.S., rue d'Ulm Paris. Résultats non publiés.

[TAR2] L. Tartar, Estimations fines de coefficients homogénéisés. Colloque en l'Honneur de E. De Giorgi. Research Notes in Mathematics, Pitman (1985).

[ZE] G. Zérah,Thèse 3e cycle Université Paris Sud.

[ZO] J.P. Zolezio, Sur le contrôle de la première valeur propre d'un opérateur elliptique variationnel d'ordre 4 par rapport aux coefficients de l'opérateur et par rapport au domaine. Preprint Université de Nice Avril (1978).

[LC] K.A. Lurie - A.V. Cherkaev, Optimal structural design and relaxed controls. Optimal control Appl. Meth. vol 4 (1983).

R. Tahraoui
Université de Paris-Sud
Centre Universitaire d'Orsay
Département de Mathématiques
Bâtiment 425
90405 ORSAY
France
et
Université de Picardie
U.F.R. maths. Info.
33, rue Saint Leu
Amiens 80039
France

M I WEINSTEIN

Ground states of nonlinear dispersive equations: stability and singularities

0. Introduction

In this paper I survey some results concerning certain nonlinear bound states for two nonlinear dispersive equations. These are the nonlinear Schrödinger equation

$$\text{(NLS)} \quad i\phi_t+\Delta\phi+f(|\phi|^2)\phi = 0 \qquad \phi\in C([0,\infty);H^1(\mathbb{R}^N))$$

$$\phi(x,0) = \phi_0(x)\in H^1(\mathbb{R}^N)$$

and the generalized Korteweg-de Vries equation

$$\text{(GKdV)} \quad w_t+a(w)w_x+w_{xxx} = 0 \qquad w\in C([0,\infty;H^2(\mathbb{R}^1))$$

$$w(x,0) = w_0(x)\in H^2(\mathbb{R}^N).$$

The existence theories of Ginibre and Velo [1] for NLS and Kato [2] for GKdV guarantee the existence of global solutions, in the above function spaces, provided the nonlinear interaction terms f(t) and a(t) are suitably restricted. For ease of presentation I consider the special cases $f(|\phi|^2) = |\phi|^{2\sigma}$ and $a(w) = (2\sigma+1)w^{2\sigma}$, where $\sigma<2/N$ is sufficient for global existence. Throughout I assume $\sigma\leqslant 2/N$, and refer to $\sigma = 2/N$ as the critial case.

1. Solitary waves

As models of nonlinear wave propagation, NLS and GKdV have localized, finite energy solutions which we call nonlinear bound states, or solitary waves. We seek these solutions in the form

$$\psi(x,t) = u(x,E)e^{iEt} \quad \text{for NLS, and} \tag{1.1}$$

$$\psi(x,t) = u(x-Et) \quad \text{for GKdV.} \tag{1.2}$$

Substitution of these wave forms into NLS and GKdV yields a semilinear elliptic problem for the function u

$$\Delta u - Eu + |u|^{2\sigma} u = 0 \qquad u \in H^1(\mathbb{R}^N) \tag{1.3}$$

In dimensions $N > 1$, (1.3) has infinitely many solutions. Of particular interest is the ground state solitary wave, a positive, radial, monotonically decreasing solution of (1.3). The existence of solutions to (1.3) has been proved in [3,4], for the range $0 < \sigma < 2/(N-2)$. Uniqueness has been shown for some subrange of σ-values [5,6], though it is believed to be true for the entire range of σ.

When $N = 1$, $\sigma = 1$ for NLS and $N = 1$, $\sigma = 1/2$ or 1 for GKdV, these equations are known to be solvable by the inverse scattering transform (see, for example, [7]). Here, the ground state solitary wave is the "one-soliton", which has remarkable stability properties, and which is known to play an important role in the structure of general solutions of the initial value problem.

2. Stability of ground states

Here, I will sketch a new proof of orbital stability of ground states, provided $\sigma < 2/N$. We use the Lyapunov method. The additional restriction $N = 1$ or $N = 3$ is required due to a technical point connected with uniqueness of ground states which we conjecture to hold for all σ and N with $0 < \sigma < 2/(N-2)$. A detailed presentation appears in [8]. Stability for NLS was proved by compactness methods in [9]. Stability for KdV ($a(t) = t$) was proved in [10,11]. Results on the stability of Langmuir waves and generalizations of KdV have been obtained independently in [12,13].

I shall first discuss the notion of stability used. Note that if $\phi(x,t)$ solves NLS and $w(x,t)$ solves GKdV, then $\phi(x+x_0,t)e^{i\gamma}$ solves NLS and $w(x+x_0,t)$ solves GKdV for all $x_0 \in R^N$ and $\gamma \in [0,2\pi)$. The orbit of a ground state $R(.,E)$ is defined to be

$$\begin{aligned} G_R &\equiv \{R(.+x_0,E)e^{i\gamma} \mid x_0 \in \mathbb{R}^N,\ \gamma \in [0,2\pi)\} \quad \text{for NLS, and} \\ &\equiv \{R(.+x_0,E) \mid x_0 \in \mathbb{R}^1\} \quad \text{for GKdV.} \end{aligned} \tag{2.1}$$

The deviation of the solution $\phi(x,t)$ from G_R is defined by

$$[\rho_E(\phi(t),G_R)]^2 \equiv \inf \{\|\nabla\phi(.+x_0,t)e^{i\gamma}-\nabla R(.,E)\|^2 + +E\|\phi(.x_0,t)e^{i\gamma}-R(.,E)\|^2\}.$$

Here, the infimum is taken over $x_0 \in \mathbb{R}^N$ and $\gamma \in [0,2\pi)$ for NLS and $x_0 \in \mathbb{R}$ with $\gamma \equiv 0$ for GKdV.

In the special case of the power nonlinearity the stability result is :

THEOREM 1. Let $\sigma < 2/N$ with N = 1 or N = 3. Consider the initial value problem for NLS or GKdV with initial data f. For every $\varepsilon > 0$, there is a $\delta(\varepsilon) > 0$, such that if $\rho_E(f,G_R) < \delta(\varepsilon)$, then for all $t > 0$, $\rho_E(\phi(t),G_R) < \varepsilon$ for NLS ($\rho_E(w(t),G_R) < \varepsilon$ for GKdV).

The theorem is proved by constructing a Lyapunov functional using conserved integrals of NLS or GKdV. The conserved integrals employed are

$$N(\psi) = \int|\psi|^2 dx \quad \text{and} \tag{2.2}$$

$$H(\psi) = \int|\nabla\psi|^2 - \frac{1}{\sigma+1}|\psi|^{2\sigma+2}dx. \tag{2.3}$$

Our Lyapunov functional is

$$E[\psi] = H[\psi] + EN[\psi]. \tag{2.4}$$

The key point is to show that, restricted to functions with L_2 norm equal to $\|R(.E)\|_2$, $R(.,E)$, modulo translations (and phase), i.e. in the metric ρ_E, is a local minimum of E provided

$$\frac{d}{dE} N[R(.,E)] = -\frac{1}{E}\frac{d}{dE} H[R(.,E)] > 0. \tag{2.5}$$

Stability of R(;E) under condition (2.5) then follows. For the case $f(|\phi|^2) = |\phi|^{2\sigma}$, (2.5) is equivalent to $\sigma < 2/N$.

Condition (2.5) is the analogue of the convexity condition obtained by Shatah [20] for stability of standing waves of nonlinear Klein-Gordon equations.

3. Singularity at criticality

In the case $\sigma \geq 2/N$ ground states of NLS are not stable [14, 15, 16], in the above sense. In fact, initial data arbitrarily close to a ground state may give rise to a solution $\phi(t)$ for which $\|\nabla\phi(t)\|_2 \to \infty$ as $t \to T < \infty$, for some $T > 0$. In the critical case $\sigma = \frac{2}{N}$, however, the ground state does appear to play a stable role in the asymptotics of blowing up solutions. Numerical experiments in the critical cases $\sigma = 1$, $N = 2$ and $\sigma = 2$, $N = 1$ [17, 18] indicate for a large class of initial data that the singular solution has the form

$$|\phi(x,t)| \sim [\, a(t)]^{-\frac{N}{2}} R\left(\frac{x}{a(t)}\right) \tag{3.1}$$

where R is a ground state of (1.3) and $a(t) \to 0$ as $t \to T$.

Toward an understanding of the observed asymptotics (3.1), I have proved the following result. I will assume that the ground state is unique. This has been proved for $N < 4$ [5,6].

Let S_λ denote the dilation scaling : $(S_\lambda\phi)(x,t) = \lambda^{\frac{1}{\sigma}}\phi(\lambda x,t)$.

THEOREM 2. Let $\sigma = 2/N$ and

$$\|\phi_0\|_2 = \|R\|_2 \quad . \tag{3.2}$$

Assume that the initial value problem for NLS with initial data ϕ_0 has a solution that blows up, i.e. $0 < T \leq \infty$ such that $\lim_{t\to T} \|\nabla\phi(t)\|_2 = \infty$.

Then, there is a function $\lambda(t) \to 0$ as $t \to T$, such that for any sequence $t_k \to T$,

$$S_{\lambda(t_{k_j})}\phi(t_{k_j}) \to R(.) \tag{3.3}$$

(up to translations in space and phase) strongly in H^1.

The proof uses the fact that a suitable time-dependent scaling of a solution as $t \to T$ forms a minimizing sequence of a particular variational problem. Convergence of a subsequence, is studied within the framework of the *concentration compactness* principle [21]. A detailed presentation appears in [19].

The critical *radius* $||R||_2$ has arisen in previous work on NLS at criticality. In [15] , I showed that the ground state is a threshold in the following sense.

<u>THEOREM 3</u>. Let $\sigma = 2/N$ and $\phi_0 \in H^1$. If

$$||\phi_0||_2 < ||R||_2 \tag{3.4}$$

then NLS has a unique global solution of class $C([0,\infty);H^1)$.

Condition (3.4) is sharp in the following sense.

<u>THEOREM 4</u>. There exist initial data ϕ_0 with $||\phi_0||_2 = ||R||_2$ for which the solution of NLS blows up in finite time.

<u>Proof</u>. To show this we use the existence of a new symmetry of NLS when $\sigma = 2/N$. Namely, if $\phi(x,t)$ solves NLS, then for any a,b,c and d satisfying $ad-bc = 1$,

$$\frac{1}{(a+bt)^{N/2}} \phi \left(\frac{x}{a+bt}, \frac{c+dt}{a+bt}\right) \exp \frac{ib|x|^2}{4(a+bt)}$$

solves NLS.

Application of this symmetry to the exact solution $R(x)e^{it}$ with a and b chosen so that $ab < 0$, yields the result. □

Finally, the situation for GKdV in the critical case, i.e. $a(w) = 5w^4$, appears to be analogous. It is believed that there exist solutions of GKdV in the critical case, which blow up in finite time, though a proof of this conjecture has been elusive. Assuming this to be the case, we can show that Theorem 2 holds for GKdV as well, with simple modifications [19].

References

[1] Ginibre J. and Velo G., On a class of nonlinear Schrödinger equations I, The Cauchy problem, general case. J. Func. Anal. 32 (1979), p.33-71.

[2] Kato T., On the Cauchy problem for the (generalized) Korteweg-de Vries equations. In Studies in Appl. Math. Advanced in Mathematics Supplementary Studies 8 Academic Press (1983).

[3] Strauss W.A., Existence of solitary waves in higher dimensions. Comm. Math. Phys. 55 (1977), 149-162.

[4] Berestycki J., Lions P.L., Nonlinear scalar field equations I, II Existence of a ground state. Arch. Rat. Mech. Anal. 82 (1983) p.313-376.

[5] Coffman C.V., Uniqueness of the ground state solution for $\Delta u-u+u^3 = 0$ and a variational characterization of other solutions. Arch. Rat. Mech. Anal. 46 (1972) p. 81-95.

[6] McLeod K. and Serrin J., Uniqueness of solutions of semilinear Poisson equations. Proc. Nat. Aca. Sci. USA 78 (1981) p. 6592-6595.

[7] Whitham G.B., Linear and Nonlinear Waves. John Wiley and Sons. New York (1974).

[8] Weinstein M.I., Lyapunov stability of ground states of nonlinear dispersive evolution equations. Commun. Pure Appl. Math., to appear.

[9] Cazenave T. and Lions P.L., Orbital stability of standing waves for some nonlinear Schrödinger equations. Comm. Math. Phys. 85 (1982) 549-561.

[10] Benjamin T.B., The stability of solitary waves. Proc. Roy. Soc. Lond. A 328 (1972 153-183.

[11] Bona J., On the stability of solitary waves. Proc. R. Soc. Lond. A 344 (1975) 363-374.

[12] Laedke E.W. and Spatschek K.H., Stable three-dimensional envelope solitons. Phys. Rev. Lett. 52 (1984) p.279-282.

[13] Laedke E.W. and Spatschek K.H., Stability theorem for KdV-type equations. J. Plasma Physics, to appear.

[14] Berestycki H., Cazenave T., Instabilité des étas stationnaires dans les équations de Schrödinger et des Klein-Gordon nonlinéaires. C.R. Acad. Sc. 293 (1981) p. 489-492.

[15] Weinstein M.I., Nonlinear Schrödinger equations and sharp interpolation estimates. Comm. Math. Phys. 87 (1983) p. 567-576.

[16] Shatah J. and Strauss W., Instability of nonlinear bound states, Comm. Math. Phys., 100 (1985), p. 173-190.

[17] Zakharov V.E. and Synakh V.S., The nature of the self-focusing singularity. Sov. Phys. JETP 41 (1976) p. 465-468.

[18] Sulem P.L., Sulem C. and Patera A., Numerical simulation of singular solutions to the two-dimensional cubic Schrödinger equation. Commun. Pure Appl. Math. 37 (1984) p. 755-778.

[19] Weinstein M.I., On the structure and formation of singularities in solutions of nonliear dispersive equations.Comm.PDE,11(1986), 545-565.

[20] Shatah J., Stable standing waves of nonlinear Klein-Gordon equations. Comm. Math. Phys. 91 (1983) p. 313-327.

[21] Lions P.L., The concentration-compactness principle in the calculus of variations : the locally compact case, parts 1,2 . Ann. Inst. Henri Poincaré, Analyse Non Lineaire, 1(1984), 109-145 and 223-283.

Michael I. Weinstein
Princeton University
Department of Mathematics
Fine Hall, Washington Road
Princeton N7 08544
U.S.A.